Get 10,000+ Twitter Followers

Easily, Quickly, Ethically

STEP-BY-STEP: YOU CAN DO IT!

Cathy Turney

Get 10,000+ Twitter Followers
Easily, Quickly, Ethically
Step by Step: You Can Do It!

Cathy Turney

Published by: Real Estate Success Press
www.RealEstateSuccessPress.com

ISBN: 978-0-9982138-0-4

Trademark disclaimer: Instead of inserting a trademark symbol every time a trademarked name is listed, we have used these names in an editorial fashion with no intention of infringement on the trademark. Where these trademarked names appear in this book they have been printed with initial caps. All trademarks belong to their respective owners.

Limits of liability: Methods to implement the tools in this manual were accurate at the time of writing; however, software applications continually undergo modification and price changes. No warranty is given by the author or publisher regarding the information in this book, nor is any liability assumed by the author or publisher for any errors or omissions or damages consequential to the application of this information.

Author's Note

Do you not "get" Twitter but know you must? Besides being the No. 1 place to promote your product or service, Twitter has become the de facto news source today.

For Twitter to be enjoyable and productive, it is important that you stay current with its constantly evolving platform, not to mention the resulting changes in the various supporting apps and programs. This book is in its third update. I strive to make it the only Twitter how-to that you'll ever need!

Endorsements

“I've had a Twitter account for years and actually have almost 7,000 followers. But at the end of the day, having a bunch of followers does not ensure success or relationship-building, which is what social media is all about. Cathy Turney has created a guide that is almost like a conversation we're having over a good glass of wine…and as a result I'm seeing all the incredible potential of this platform. Her humor makes it all the more fun to read. While I am strongly recommending it for real estate agents, anyone can benefit from the infinite possibilities that Twitter and this book have to offer.”

—Tamara Dorris, real estate agent, author, instructor

“As a book-writing and publishing coach, I am recommending this book to my clients! Most of them need to reach more prospective readers—before their book launch or even before sending out a book proposal. I’ve recommended the book to several clients and since then they are already finding great results in growing their Twitter following. It’s exciting to see how quickly the book has helped them.”

—Lisa Tener, Stevie Award-winning book coach, author,
faculty member for Harvard Medical School CME publishing course

“Cathy Turney’s book on using Twitter and gaining followers is great! It's also really enjoyable to read because, besides presenting the most important information about Twitter in an easy-to-understand 8.5” x 11” format with lots of screen-shots, it’s also full of bird puns! It’s a bird’s-eye view of only the most important concepts you’ll need to take your Twitter skills to the "nest" level. If you’ve been feeling "hawkward" about your ability to gain followers and fly confidently through the Twitterverse, I hope you’ll read (and enjoy) Cathy’s book as much as I did. "Hoot" have thought it could be this easy?!”

—Dr. Lynelle Schneeberg, Psy.D.

“I wanted to learn more about how to use this social media phenomenon that is sweeping the world. And I did, thanks to the author's informative and entertaining guide. I’ve had a Twitter account for years but haven’t done much with it as a marketing tool. Cathy Turney approaches what could be an intimidating challenge with extensive research and experience spiced with a disarming sense of humor that will send you soaring into the Twittersphere.”

—Robert L. Haught, author

"Social media is important. As business owners, we understand that. So, why is it that so few business owners use social media effectively? In part, it's because using social media is a skill—and just like with any learnable skill, the student will become accomplished only if she or he is in front of the right teacher. With "Get 10,000+ Twitter Followers," business owners now have a teacher (Cathy Turney) who not only knows how to turn them into Twitter experts but makes the process enjoyable and fun at the same time. We plan to apply many of the methods in this book to the marketing efforts in our business…so that our own Twitter account will finally take flight!"

—John Robinson, founder of PassionQuest Technologies, LLC,
No. 1 best-selling author, and master business coach

Foreword

Let's be honest. Are you famous enough to sell your product or service as effectively as you want? Are you targeting the consumers who are most likely to love what you do? *Does Twitter intimidate you?*

You are not alone. Cathy Turney's book will get you past that anxiety and over the learning curve so you can become one of the Twitterati. And you'll laugh all the way to the bank.

I met Cathy at a National Society of Newspaper Columnists conference, where she presented a segment on self-publishing. Who was this person who openly admitted before a crowd of communicators that she's a technophobe? That anything beyond Facebook was beyond her? That she got dragged kicking and screaming onto Twitter? A very smart cookie who, in spite of said tech challenges, self-published her way to a No. 1 best seller on Amazon and won the American Business Awards Gold Stevie nationally and internationally for *Laugh Your Way to Real Estate Sales Success*, her humorous inside story on the real estate sales business.

Cathy sincerely wants others to succeed. She overcame her Twitter phobia to get far beyond 10,000+ followers, and she now shares what she's learned from her journey in this laugh-out-loud tutorial to help others like her—right-brain creatives and Luddites as well as the technologically gifted.

Is Twitter important to your success? Will this book get you to 10,000+ Twitter followers easily, quickly, and ethically? I answer with a resounding "yes!"

—Paul Lander; 42,000+ Twitter followers; award-winning writer and producer for Showtime, the Disney Channel, ABC Family, XM/Sirius, Lifetime, The American Bystander, McSweeney's, Humor Times, *and* National Lampoon

Table of Contents

About This Book

What does anyone with a product or service to promote need?

A way into millions of minds, hearts, and pocketbooks.

What did the 2016 U.S. presidential election teach us?

That Twitter is the way into millions of minds, hearts, and pocketbooks.

Twitter is the fastest, easiest, most fun, least expensive (*need I say more?*) method to see and be seen by tens of thousands of people who want to hear what you have to say—your target audience. You and your product need to look "relevant" and having a huge Twitter following accomplishes that.

But don't panic! This book will make it E.A.S.Y.

It's like Christmas every morning for me when I go to my computer, log on, and find that I've added 25 to 50 followers (or more) over the past 24 hours, having spent only 15 minutes (or less) "trawling."

But it didn't begin like that.

A few years ago when I asked my social media guru how many Twitter followers I should have, she said, "I'd like to see you at 10,000 within the year." **!** I was at 200, and that had been a struggle. I would follow 100 people a day, spend at least an hour trying to "engage" them and, if I was lucky, five would follow me back. Worse, sometimes I was "unfollowed" and my net actually decreased. After all that effort.

So, I did what every writer does so well—I researched. I watched YouTube webinars where darling little millennials would calmly (*calmly!*) explain (rapidly) how to work Twitter, and the information was so *un*intuitive to me, and the screens would fly by so quickly, that I couldn't print them out to review later. On the fifth replay I just gave up on videos.

I bought how-to books and studied. But being a right-brain Luddite born under a technology-crossed star, I was only confused by their instructions. And what I saw on the computer screen didn't match their descriptions. They'd tell me what to do, but not explicitly *how* to do it (because they were written by *left*-brain techies). Their directions took me down rabbit holes where, I learned later after weeks of foraging through the thicket, I didn't need to go in my pursuit of how to *JUST DO TWITTER*, enjoy it, and amass a following that would make me and my own books look relevant.

Read on, my little Twitter fledgling. This book speaks to the rest of us. Soon you will soar with the big birds!

How to Use This Book

When I attend meetings I'm always amazed at how few people take notes. I'm not surprised, though, that those folks are usually the ones who remember what they're supposed to do. I suggest that if you are at this moment reading the electronic version of this book, you might want to purchase the companion softcover edition, too. I've already "taken notes" for you. And vice-versa, if you're just reading the softcover edition, which is black and white, it won't be as easy to distinguish between multicolored phrases and words that comprise the ~~alphabet soup~~ different components of tweets as they appear on the computer screen. Having both the online and paper copy versions of the book open to the corresponding pages will make learning a lot easier.

This tutorial is conducted on a laptop—not a handheld device. It's easier to begin on a big screen with a big keyboard, and we'll all be on the same "page." You can do lots of things on your phone that will keep you highly entertained—later. But for learning these techniques you'll prefer the desk/laptop. And, frankly, it's where I prefer to roost when spending five to ten minutes a day gathering my flock.

It is my sincerest hope that the screens I show you match what you'll see on your own. But please bear in mind that technocrats are famous for "improving" the experience and our screens may not match exactly. The words that instruct you to take a particular action should appear somewhere on your screen. And seriously—if you get stuck I will help you! See "About the Author, and…" at the end of this book for how to reach me. I want you to succeed.

Caveat

Part I of this book is profoundly simplistic. Don't let that turn you off—it's written for the most trepidatious wannabe tweeter. If you already have a Twitter account, know how to read and interpret other people's tweets and understand the lay of the land, I would still advise reading Part I because it has tips on enriching the Twitter experience. Part II is where I explain in ~~excruciating~~ detail, with ~~screenshots~~ pictures, exactly how I find and gather my flock—the techniques to streamline and speed your progress.

Much of my advice runs counter to what you may have read in other books. But that's because amassing a huge following *does not have to be that hard!* Always remember our objective: to "easily, quickly, and ethically" increase your following. There is no right way or wrong way to do Twitter, but I believe this is the best way.

I am going to show you how to love Twitter. Let's begin!

PART I – Feather Your Nest for Launch into the Twittersphere

As mentioned, Part I may be excruciatingly simple to you, but when I was signing up for Twitter, I kept wondering if the instructor and I were on the same page, literally. I don't use screenshots for the account setup section because Twitter is famous for changing the appearance of pages when its workers get bored and then you'll be even more confused. I guarantee, though, that you will find the words in these instructions somewhere on the pages I direct you to. If something doesn't match up, just take a deep breath, relax, and read. The answer will be there because, rest assured, Twitter wants (and needs) you to sign up.

In Part I we'll lay the groundwork for launch:

- Setting up your skeleton account—that's all you need to start having fun
- The three keys to unlock any tweet (it's soooo easy)
- "Voyeurism," or searching interesting topics

Twitter does dozens of wonderful things, but you don't need to know about or do most of them to accomplish your goal of being relevant and exposing your work to the world.

Skeleton Account Setup

Good news! Twitter has greatly simplified the setup process. So, positioning yourself before your computer screen, type into your browser search bar: Twitter.com (or https://twitter.com) and choose to "Sign up." The resulting screens may not appear exactly as follows; it all depends on the phase of the moon, how their stock is doing, and other seemingly unrelated matters. If Twitter asks you if you "want this website to remember you at this address?" (or similar verbiage), check the box—it's a time-saver during your morning ritual. Here are pointers about how to fill in the information:

- At "Enter your name," use your real name. It can be up to 50 characters long, including spaces. If you're a professional with "letters after your name," including them can boost your find-ability. When I wanted to follow Diane Radford, who wrote a delightful book entitled *Bless Your Little Cotton Socks*, I did a Twitter search under that name and pulled up two Dianes. Because "my" Diane Radford included MD in her name, I could choose the correct one. Similarly, Dr. Craig Malkin, author of the most timely book *Rethinking Narcissism*, identifies himself with "Dr.," which is very good because there were two Craig Malkins on Twitter, last time I looked. Having said all that, Twitter will create your "handle" from your name. If your name exceeds 15 characters, they'll abbreviate it. You can change it later, but let's not go there now.
- "Enter your phone" must be a cell phone number because Twitter will send you a confirmation code via text message. You will type in that code on your computer when asked for it during the setup process.
- Enter your DOB (date of birth).
- Skip past "Customize your experience."
- At "Create your account," click "Sign-up."
- "Verify phone" —click "OK" and enter the code they text you.
- "You'll need a password." It can be a minimum of six characters.
- "Pick a profile picture." Either upload your image (I advise making it the real you because people are more likely to follow real people), or "skip for now."
- "Describe yourself." You get 160 characters. If you're stumped, check out what others have written for ideas. When finished, click "Next."
- "What are you interested in?" Choose a few, then click "Next."
- "Suggestions for you to follow." You can choose from this list or not. When done, click the blue jelly bean (upper right) that says "Follow."

You have just set up your account! Welcome to the aviary!

Now, let's log out and log back in again. But where? Twitter hides this. At the top of the page you will see a search bar (the area with the little magnifying glass) and a circle with your image in it. Click on your image. This column will appear:

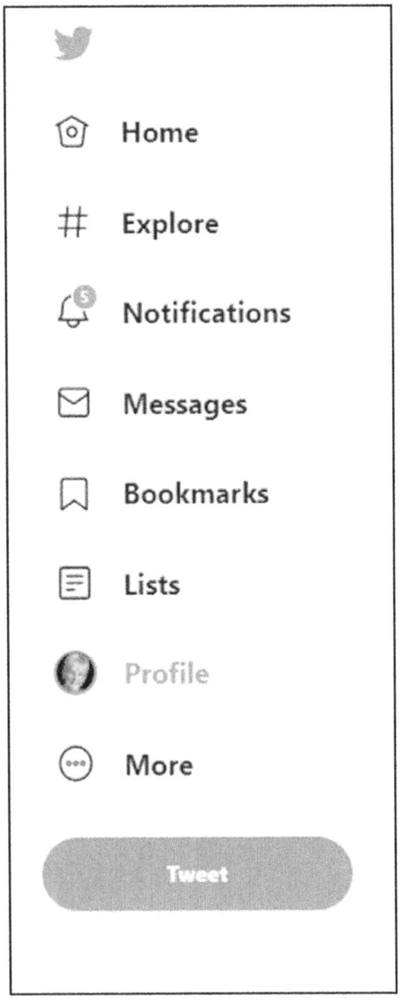

Click on "More," and "Logout" will be at the bottom of that list.

That will take you back to the sign-in screen. Type in your name and password, then click "Login."

That column to the left will reappear. Click on "Profile." Under your image you'll see your name, and under that will be your "handle," which Twitter created for you and is limited to 15 characters. It begins with "@." If you don't care for it, there's a way to change it, but let's not go there yet.

Next, we'll nest a while and I'll show you how to interpret tweets.

The Three Keys to Interpreting Any Tweet

They are:

- # - hashtag
- @ - "at" sign, aka "handle"
- Web links

The simplest way to understand what's going on in tweets is this: Instead of addressing someone as Mr. Tim Brown or Ms. Jane Jones, in Twitter they'd be @TimBrown and @JaneJones. The "at" sign (@) is just their "Mr." or "Ms." About the hashtag: Pretend you're at a party and want to find an interesting conversation to join. Imagine placards over the clusters of people indicating what they're discussing—#humor or #kids would be examples. You would select a group to converse with by finding a hashtag that denotes the topic you're interested in. That's how it works on Twitter.

Now, back at your computer, let's log in to Twitter again by typing in https://twitter.com. You'll be asked your username and password. After typing that in you will land on a screen that says "What? No tweets yet?" Or you'll see a stream of tweets posted by your followers if you have any yet. (It's rare that there won't be any there soon; Twitter seems to suggest people for other newbies to follow.) These tweets are updated second by second. The first tweet that popped up for me (the result of Providence and being a dog lover) was this adorable image:

Now remember, at this point we're not learning how to search Twitter to find things. No. We're just learning how to ~~interpret the jibberish that constitutes a Twitter post~~ read a tweet. One easy thing at a time. Focus.

Let's break this down. The top line of a tweet almost always consists of the "avatar" (the face) that identifies the tweeter, his or her "handle" (Twitter name preceded by @), and the date the tweeter posted.

In the top line of this tweet we have:

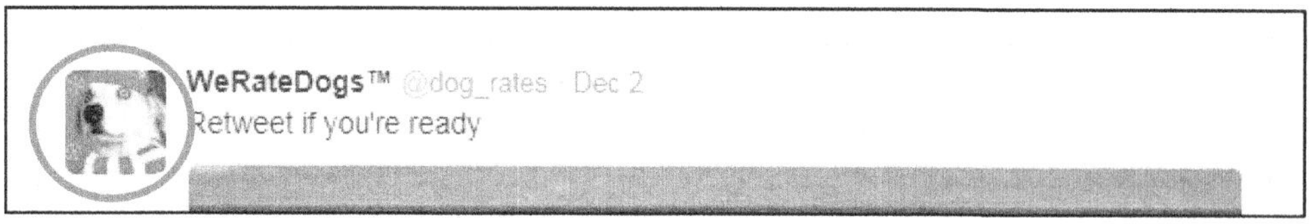

The image of the dog, which I've circled above, is the avatar. Usually tweeters use a photo of themselves, but a business is posting this tweet, so it used some poetic license.

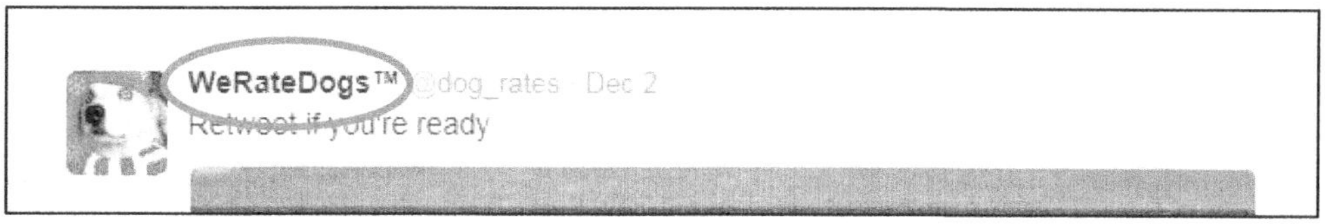

"WeRateDogs," circled above, is the tweeter's company name.

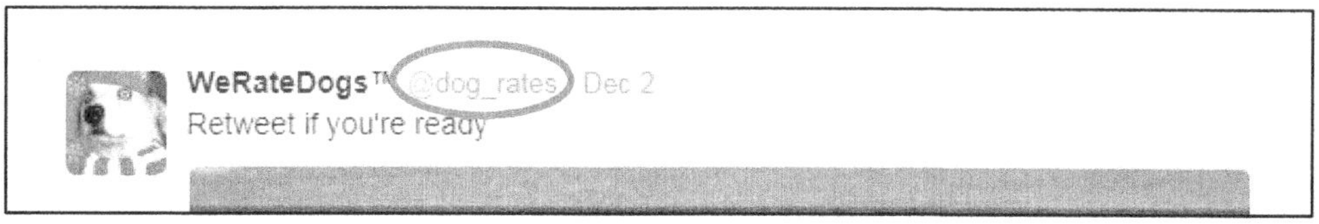

"@dog_rates" (above) is the tweeter's "handle," his Twitter name.

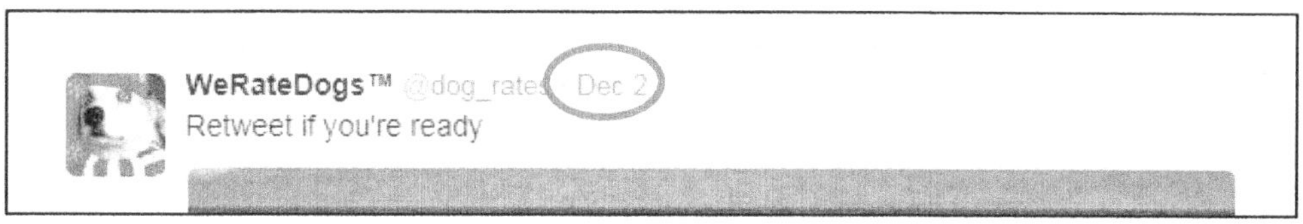

"Dec 2" (above) is the date this tweet was posted

"Retweet if you're ready" is the message—short and sweet.

Below that top bar is the adorable dog photo the company posted which, if you were online and clicked on it, would show you a larger size photo of same:

Underneath the photo are these symbols (and they are pretty much all the same in every tweet you'll see):

 This is the "Reply" button. Clicking here will take you to a screen where you can type a tweet to the person whose tweet you're commenting on. Replies are *not* private (vs. a "message," below). When you exceed 280 characters, the excessive characters will turn red, letting you know that you need to shorten it.

 This is the "Retweet" button. It's saying that this tweet was retweeted twice. If you click there, you can spread the word, too.

 You'll recognize the heart here. Four people loved this tweet.

This is the “Direct Message” spot, private communiqués to the tweeter and has no character limit.

This is some Twitter analytics stuff which is not critical to understanding tweets and is rather ambiguous, so we won’t cover it here.

Now that you know what those little figures mean, let’s decipher a few more tweets. It’s so easy!

Joanna Penn is a prolific self-publishing and marketing guru (www.thecreativepenn.com). Per her Facebook “About” section, “Joanna Penn is a *New York Times* and *USA Today* best-selling author of thrillers. Her site, TheCreativePenn.com, is regularly voted one of the Top 10 sites for writers.” Definitely someone whose tweets are worthy of interpreting. Here’s one, followed by an explanation of what she’s doing:

Line 1: 

This means that Joanna “pinned” this post to the top of her tweet stream so that it will remain as the first one that readers see—until she unpins it or pins another one. We’ll go into pinning later in the book.

Line 2: 

This is Joanna Penn's real name, followed by her handle (Twitter identifier).

Line 2: 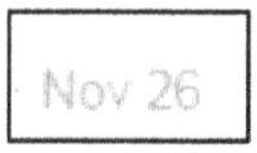

is the date she posted this tweet.

Line 3: 

"#writing" puts this tweet in the Twitter feed (or, in Facebook terms, "stream") of anyone else who is following the topic #writing. The hashtag simply acts as a congregator of conversations on Twitter. Like at a party—this group of attendees would be chatting about writing; a #doglover cluster would be posting cute dog pictures and stories; a #Disneyland group would give you several chat groups where you could learn up-to-the-minute things going on at Disneyland.

Line 6: 

This is an electronic link to one of her web pages where you can read about and buy her book. (Seeing it online, in a color different from the text of the tweet, makes it more obvious that it's a link.)

If you were to place your cursor anywhere in the image above and click, it would give you a larger image of same. There are two things that can happen when you click on the main image (if one is posted): you'll be shown either a larger picture of that image or you'll be taken to a website where more information about it is posted.

See how easy that was?

Let's decipher another tweet, so you can see that this is really all there is to it!

CL Bennett is hugely popular in the #amwriting "collective" where zillions of authors and writers post using that hashtag/address. It's a place where you can let others know about your writing experience and get support. Here's one of her tweets followed by some explanations:

Line 2: #amwriting By including "#am writing," CL channels the post into its own conversation stream where others posting with the same characters or viewing that stream will see it.

Line 2: #SundayFunday She also hashtag-funnels this tweet into the stream for SundayFunday.

Line 3: #share Likewise, the hashtag directs this tweet to everyone following "#share."

And again, if you just click on the photo (when you're actually on Twitter) you'll see a larger version of the photograph.

Some tweets can look like alphabet soup—until you hover over the words. For instance, the gobbledygook circled below:

If you were to hover (on Twitter, not here) over this part: @MMDragons_Blade (not over buff.ly/2kZXqKD) and click on it, you would be taken to Michael R. Miller's Twitter page and you would see his own tweets plus all the tweets that mentioned him

(@MMDragons_Blade). If you were to hover over buff.ly/2kZXqKD, you would be taken to a web page where his book, *The Dragon's Blade*, was reviewed.

Pretty straightforward, right?

Apropos of alphabet soup, it's not a good idea to list a lot of @addresses unless you're thanking them for following you, as this one does:

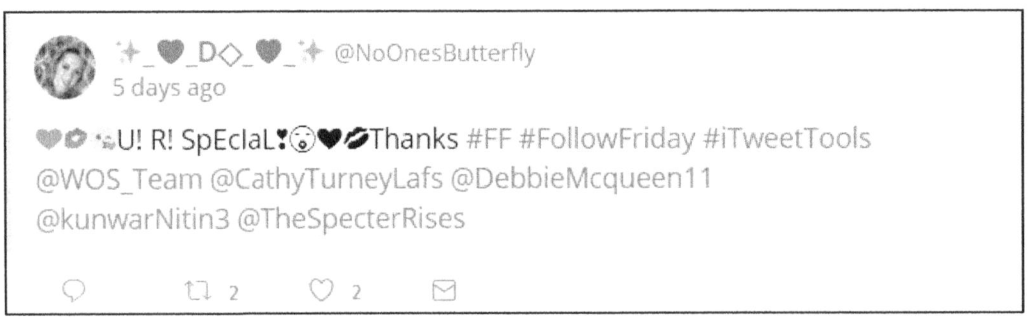

Line 1: ✨_♥_D◇_♥_✨ Instead of using her real name, this tweeter decided to use graphics. Cute, although it makes it really hard for anyone who doesn't know what her @ handle is to find her.

Line 3: U! R! SpEcIaL❣♥Thanks What she's saying is that you are special. The rest is just cute graphics followed by the word "Thanks." However, all the hearts and kisses and emojis (little images) make it very confusing to follow what she's saying. Very few tweeters post emojis to that extent, so don't worry that you'll need to translate a lot. You can skip those!

Line 3: #FF #FollowFriday #iTweetTools The hashtag addresses (#FF, #FollowFriday, #iTweetTools) put this post into those streams so those followers will see her message, which is thanking everyone she "@"ed in lines 4 and 5:

@WOS_Team @CathyTurneyLafs @DebbieMcqueen11
@kunwarNitin3 @TheSpecterRises

If you get bogged down in the (infrequent) emoji overload, just skip that tweet. Your time is better spent on other newsy posts in the same stream.

You now know how to read a tweet—yes you do; that's really all there is to it. Next, let's have some fun trawling for topics so you see how "au courant" and newsworthy and just plain interesting Twitter can be!

Voyeurism—How to Search Interesting Topics

Even before a journalist's report appears in newspapers, he's probably tweeted a few things about the breaking story already. For instance, *San Francisco Chronicle* reporter Kevin Fagan posted this tweet about the Oakland, California, warehouse fire, one of the worst in recent California history, within hours of the fire igniting. His story hit the *San Francisco Chronicle* print edition front page on December 4. He had posted this tweet simultaneously with the online version of the *Chronicle*. Unless an interested party subscribed to the *Chronicle* online, she wouldn't have had this information as quickly as she could have had via Twitter. Concerned people across the nation could read the minute-by-minute story unfold on Twitter.

Let's interpret this tweet:

Line 1: Kevin Fagan @KevinChron · Dec 3 tells us that Kevin Fagan, whose handle, or Twitter name, is @KevinChron, posted this tweet on December 3.

Line 3: sfgate.com/bayarea/articl is a web link to where the whole story appears.

Line 3: via is just another word for retweet, so here Kevin acknowledges that there was prior posting about this news at @SFGate, which is the newspaper's handle, where all its tweets appear in its stream.

Line 3: #oakland fire By putting the hashtag before Oakland fire he feeds this tweet into the "conversation group" or stream where anyone wanting to learn about the Oakland fire would look.

Let's say you've created a Twitter account and would like to find out what a friend tweets. If you don't already have Twitter open on your computer, go to the browser and type in https://twitter.com. That will land you on the home page, which will look somewhat like this:

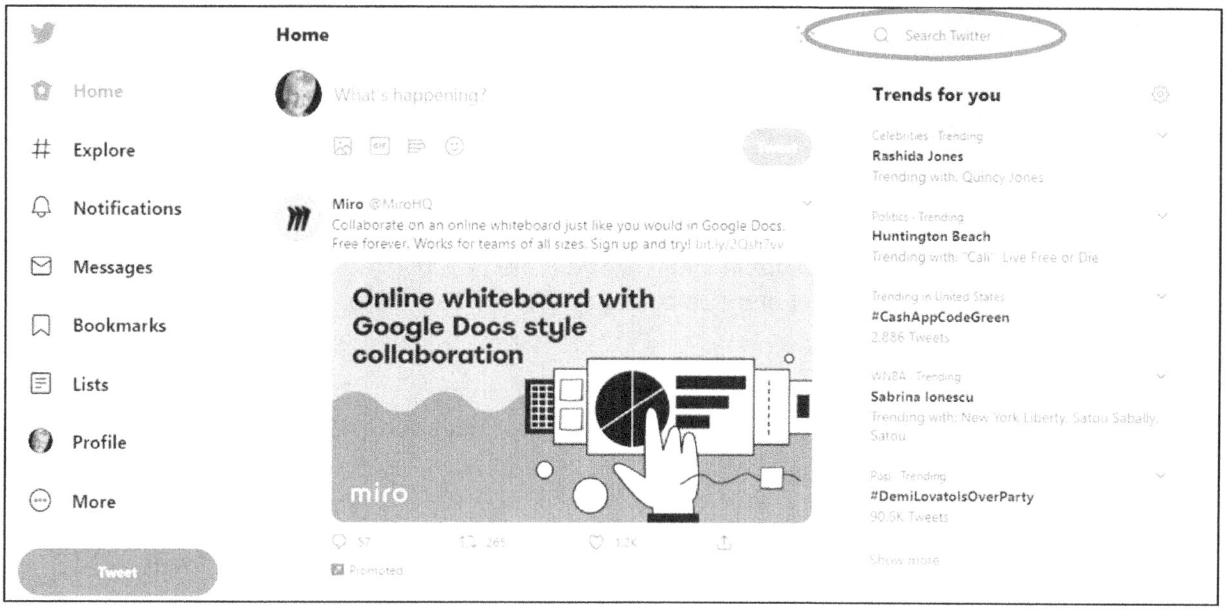

In the image above, the search box (circled) is located at the top right; it's the little magnifying glass area.

I am a huge fan of Lisa Tener (www.LisaTener.com), book writing coach. So definitely she was one of the first tweeters I wanted to follow. I knew Lisa's handle was @LisaTener, so I did a search for "@LisaTener." Lisa posts tweets packed with helpful information for writers:

Let's break it down. If you're looking at this on your computer screen, the black words are Lisa's comments, and the blue words are:

Line 3: #Books means that this tweet (which will appear in Lisa's own Twitter stream because she created it) will be posted in a feed/stream named #Books, and everyone who searches #Books will see it. Bear in mind that tweet streams are updated second by second.

Line 3: #Gifts ditto above. Everyone who follows #Gifts will see it.

Line 3: bit.ly/2hcziBV is a web link that, if you click on it, will take you to the whole article she wrote called "20+ Books That Make Awesome Gifts."

Line 3: Lisa added via and @LisaTener (on line 4), which means that this post originally appeared on her own website. Then she added @CaballoFrances @kimbersyoga @OrgArtCo because she featured their books in the article and wanted to let them know so they'll share in the Twitter love. Using the word "via" is simply how you acknowledge that others posted the same information. You don't *have* to do that when you tweet, but if you're retweeting something that another person wrote it's a courtesy that is appreciated. And it lets those people know you shared their information.

Center: The pretty image of flowers is supplied automatically from the blog post; it's the cover of Teri Scheinzeit's book entitled *Success Without Stress: Simple Steps to Finding Calm for Women Business Owners.* As mentioned before, in many cases on Twitter should you click on the image you'd just get a larger picture of it. But Lisa linked this image to the story on her website.

Here's a trending column from Huffington Post:

EASY to interpret, as are most tweets. If you were online and clicked on the link above (in the third line): huff.to/2iWZ4yv you'd land on the very cool full article with lots of pictures and inspiring quotes.

Not all tweets connect to a larger story. Some are just fun photos with a few comments. Especially in sports:

Interpreting this tweet:

Line 1: [T2W Tips and 1 other follow] means that someone with the handle T2W Tips and one other person follow JMU Football.

Line 2: [JMU Football] is the name of the entity posting, and [@JMUFootball] is that entity's handle. The tweet appeared [1h], one hour previous to my copying this image.

Line 3: The message is [Getting ready to ✈ to Texas for the 🏆!!!] and the tweeter is just using a little image of an airplane and a gold cup to substitute for text. Using hashtags in this line [#GoDukes #BeatYSU] serves to feed this post into those two conversation streams, so people following #GoDukes or #BeatYSU will see this post.

Line 4: [#StartWearingPurple] This hashtag feeds to the followers of anyone who wants to support JMU (James Madison University), which you would not know unless you

Googled "JMU" or "James Madison University," which is so way beyond what you need to know to do Twitter, so don't sweat any confusion it causes!

There will be some tweets that are just too obtuse to decipher. Move on. There will be many more in the same stream that will tell you the story. You only need to know three things: hashtag (#), handle (@), and web links, and you can get to the bottom of almost any tweet.

See how easy this is? You can speak Twitter!

One of the beautiful things about Twitter is that since you're limited to 280 characters in a tweet, none will command the whole screen and require you to page down, page down, page down until you get to the rest of the information. The web link provides that—the rest of the story. And if you don't want the rest of the story you can just scroll quickly through another 20 tweets until you find one that you do want to look into.

Now, aren't you just so excited about Twitter that your little wings are fluttering? You're almost ready to swoop in and start collecting followers.

PART II – Build Your Flock

Remember Our Objective: To Get 10,000+ Twitter Followers

It would be nice if Twitter, all by itself, as a stand-alone program, would be the end-all, be-all for gaining followers. But it so *isn't.* Like LinkedIn, it can be very obtuse. But as LinkedIn is to job-seekers, Twitter is to anyone needing to promote a product or service. You need to be there.

There are a number of programs and apps that can help you achieve the results we're after, but the following two are the ones I found to be the easiest to understand and use, most efficient, and least expensive of the bunch:

- Social Jukebox. $19.99/month. The treasure trove that provides the inspiring quotes and information tweeted out to my followers several times a day—without my lifting a finger (after initial setup).
- Buffer. They give you a free 14-day trial, and at the end allow you to continue with limited functions for free. You'll need to select the Pro ($15/month plan) to begin the trial. You don't even need this program unless you want to schedule your own personal tweets ahead of time.

When consider the monthly cost of the above, it pales in comparison to a social media expert's fees (hundreds of dollars a month) and is a small price to pay for looking relevant and marketing your product or service to thousands of targeted people. I get nothing back from these companies for recommending them, but I sing their praises at every opportunity because they make it easy for me to spend only about 15 minutes a day collecting (and keeping) dozens of followers. Use them! They'll make your professional life a lot easier.

Before I explain how to use these awesome programs, we need to do one more thing: finish setting up your profile. To this point, you have accomplished setting up your skeleton profile, which is fine if all you want to do is follow breaking news and others' tweets. But to build a flock of followers you need to look "follow-worthy," and that requires fleshing out your skeleton profile. *Which is not difficult.*

Hatch a Plan: Set Up the Rest of Your Profile

To log back in, go to https://twitter.com, enter your login information, and check the box confirming that you do want the computer to remember you. In the column at the left, click on "Profile." Then click "Edit profile' in the jelly bean under your masthead. That will yield the area below where you go in and enter the rest of your profile.

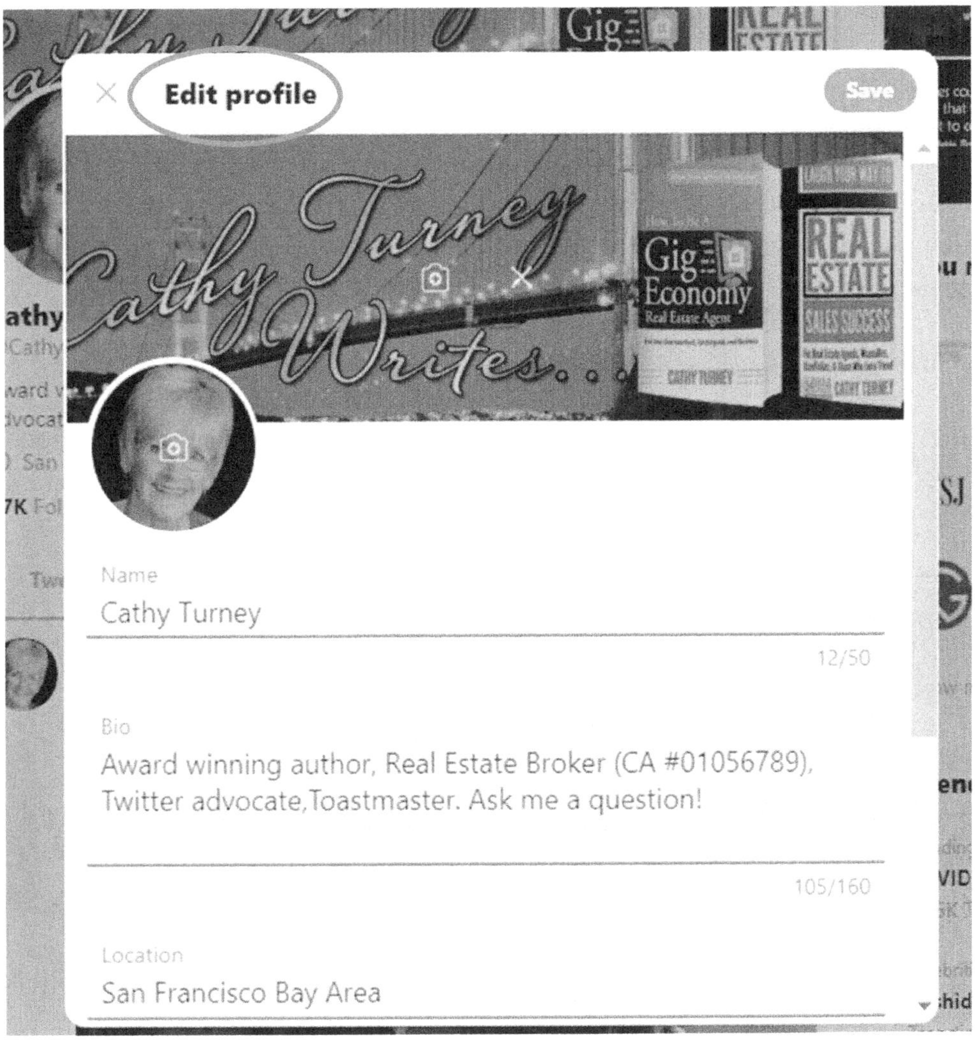

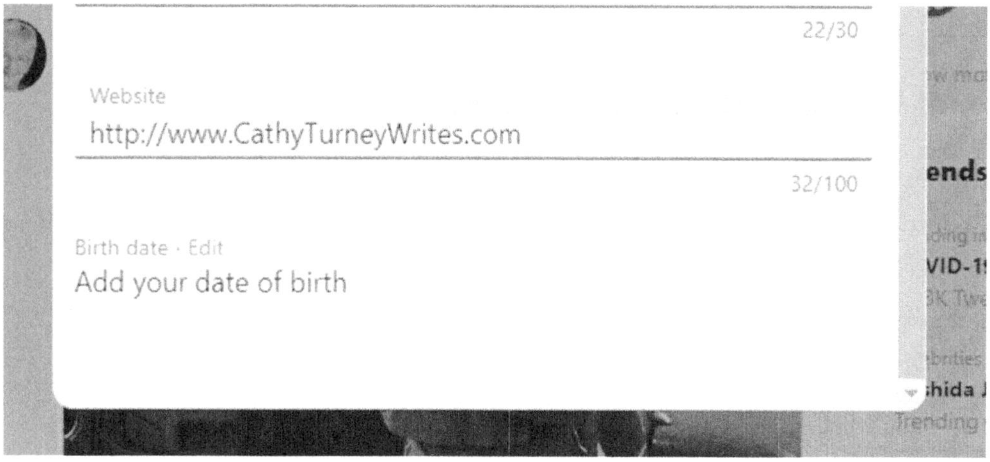

As you see, there are two areas at the top where you want to add photos or images. We need to create your masthead or banner – the colorful area at the top of your page that conveys the essence of you…which you do not need to create yourself. Below that are some questions you'll want to answer. Then press "Save."

Add Your Banner or Header

Unless you're tech-savvy, creating the perfectly sized banner for Twitter exceeds the frustration threshold set for this book. I, personally, paid a graphics person about one hundred dollars to create this:

But you can pay $25 and have Fiverr.com create a very nice one for you, perfectly sized, ready for you to store on your computer and upload in that banner area on Twitter. Here's a sample they created for me and the step-by-step instructions I filled out, which you can adapt to have a banner created for you:

I added my photo afterward in the box Twitter provides for that (as instructed above). Not bad for $25, is it? Here's how I did it:

Log on to https://www.Fiverr.com. You're asked what service you're looking for. I typed in "Twitter banner," then I hit "Search."

I scanned through the examples offered—four pages of vendors to select from. I hovered over each image. Yellow stars appear at the top of each image, their number indicating that vendor's average review rating by clients and how many jobs the vendor has done for Fiverr. Or you'll see " New Arrival," which means the vendor doesn't have a track record with Fiverr yet. However, I didn't choose from those. There's a better way.

You can narrow it down to the vendors most likely to provide a quality product by doing the following in the left column on the screen:

- Under "Delivery Time" specify three days, to give someone who is busy (and good) a long enough turnaround time.
- Under "Price Range" specify $20 to $25 (that's a competitive price for Fiverr contractors).
- Under "Seller Level" it produces this:

which indicates that for a three-day turnaround and $20 to $25 there were nine new sellers for me to choose from, no level ones, four level twos, and no top-rated sellers.

Below that, I could choose the language spoken by the contractor. I chose English.

Then, scrolling down a bit more, in the center, I clicked on “Post your request”:

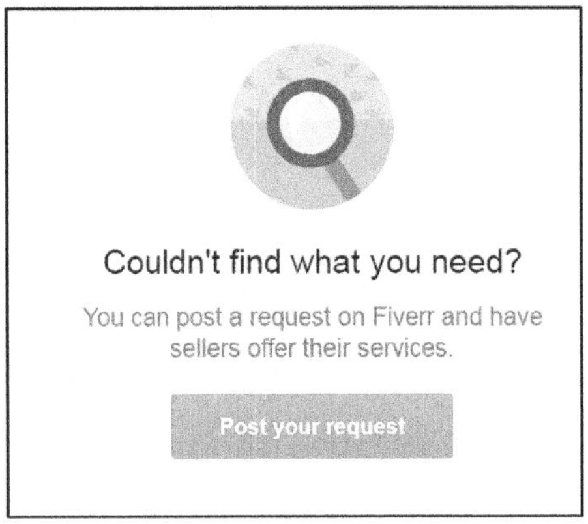

That took me to a screen where I could enter all the details about what I wanted:

- Under “Describe the service you’re looking to purchase” I entered, *I need a colorful, upbeat, Twitter banner that would attract book lovers, business people, and those wanting positive, inspiring quotes.*
- Under “Choose a category” I entered *Graphics and design.*
- Under “Subcategory” I entered *Social media design.*
- Under “Choose from the following criteria (optional)” I entered *Cover/header.*
- Under “Select Platform” I entered *Twitter.*
- Under “Once you place your order, when would you like your service delivered?” I entered *3 days.*
- Under “What is your budget for this service? (Optional)” I entered *$20.*
- Then I clicked “Post your order.”

That produced the “Manage Requests” page. Then I waited.

My first “Review offers” arrived within 24 hours via email. These are solicitations for my business per the details I’d posted, and they looked like this:

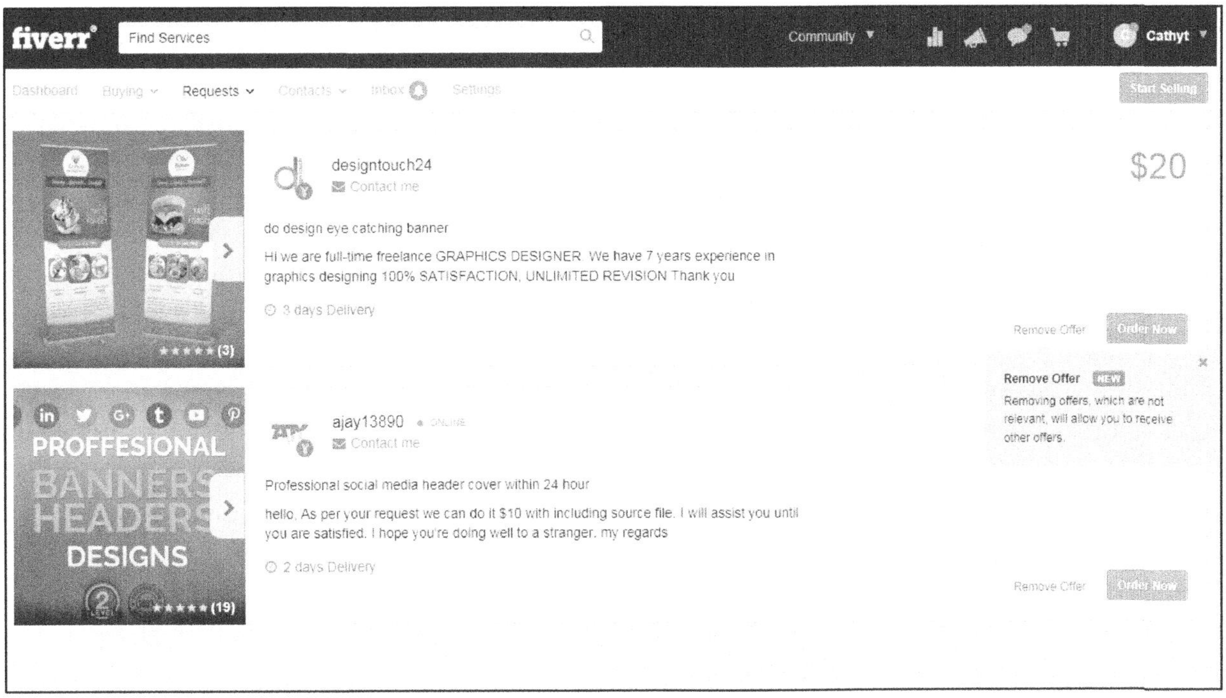

To view them, I had to log onto Fiverr and agree to pay $5 (fair enough), which was applied toward the purchase if I made one. I chose the first offer (from designtouch24), and clicked "Order Now." That directed me to the payment page, where I chose PayPal. After paying, Fiverr let me know to expect my banner within three days. It actually arrived sooner. The vendor offered to make changes for free, but I didn't need any (for this purpose). I was so delighted that I tipped him. I recommend that. These folks work hard, and frankly, would you do it for $20?!

Thoughts on Banner Content

You need to have an engaging, colorful, and upbeat banner. It's what draws followers to you when they check you out. Consider including images that reflect your hobbies, a cause, your profession. Be diverse in what you include because you'll attract followers with different interests. My banner includes an image of my book, one of the Golden Gate Bridge, which identifies my location as being in the San Francisco Bay Area, and a picture of the American Business Award that my book won. My handle, @CathyTurneyLafs, indicates that following me might be fun.

If you're stumped about what to include, go to Twitter and do a hashtag search for tweeters in your line of work (i.e., #realestate) and see what they included in their banners. Keep in mind that if you don't have cute little pictures to use, you can ask your Fiverr contractor to find some for you. Fiverr really is a one-stop-shopping resource and I've used their other services, as well, to great satisfaction several times.

Flight Instructions: The Two Easy Technology Programs to Make It Happen

Before we delve into how to upload and use each of the programs that will make you a condor among tweeters, let's discuss honesty and ethics.

There are services that advertise that they'll get you hundreds of followers if you pay them. Twitter does not like them. Twitter will shut you down if it finds you used one. Very bad.

You, however, will achieve the desired results by doing it honestly and ethically. You will serve up curated content provided by a company whose business it is to create tweets that are relevant to your topic, that followers want to read. I would almost guarantee you that any tweeter who posts good stuff three or more times a day, every day, and who also has an income-producing business or life outside of social media, has someone finding gobs of this information for them.

For instance, Guy Kawasaki (@GuyKawasaki) has 1.49 million Twitter followers and often tweets a dozen times a day. If he wrote the substance of every tweet himself he'd have no time to do anything else. He curates information from numerous sources, makes a short comment and retweets it. Followers consider him an oracle and want to have the information he deems important.

Purists may ~~sanctimoniously~~ say, "That wouldn't be the essence of me" or "It would look like I'm plagiarizing to use others' quotes." Well, first off, each quote you'll get from the service will automatically be attributed to its author, and if by chance you (on your own) lift a passage from someone else's work and quote it verbatim (in 280 characters or less), you should indicate whose words they are and, if you found it on someone else's tweet, that you are "retweeting" it. Plus! There's nothing to stop you from adding the essence of you with intermittent tweets that you create yourself. In fact, you should (*after* you get 10,000+ followers—not now: focus!).

What I'm saying here is, you may not be able to create enough interesting gems to hold thousands of followers' attention all by yourself. Why would that many people follow you just to be nice? I mean, think how much effort it takes them to lift an arm, guide a finger to the keyboard, and click "Follow." They won't do it. Only your mother loves you just because you exist. To get strangers to follow you, you must:

- Inspire
- Support
- Be fun

I know *I* can't do that three to five times a day in print. Despair no longer! Introducing ~~technology~~ the two easy programs that will do it for you. Once you're finished setting these up (which does take more than 15 minutes) you are done! You never have to do it again. You can start collecting your flock, which *does* only take 15 minutes a day—once you get the hang of it; patience is a virtue.

Before we get into how to use the two programs to collect followers, we'll set up your accounts on each of them. As a preface to the following instructions, please know that in the interest of keeping the directions as easy as possible, in most instances I have captured screenshots of the steps to take. They are the screens that appeared on the day I copied them and (breathe deeply) may have changed somewhat between then and now, as you read this. Similar information and prompts *will* be there, but may not be in exactly the same order or visual appearance. (Yoga headstand) Fear not: If you get stuck I welcome your questions and will help you. Just email me at Cathy@CathyTurney.com and put "Twitter Rescue" in the subject line.

Social Jukebox

I adore Social Jukebox. This is the program that automatically posts quotes and images for me. When I read "my" tweets, *I'm* inspired! With more than 190 collections of quotes and images by such thought leaders as Oprah, Bill Gates, and Tony Robbins and collections like "Sports Quotes," "Moods and Emotions," and "Books Go Social," you will have a hard time choosing. And the beauty of Social Jukebox is that once you set it up (instructions below), you don't have to do a thing. It automatically posts "your" words of wisdom, every day, several times a day.

As testimony to how important it is to (have someone else) post frequently, here's one tweet I received: "I noticed Cathy Turney tweets every day. She might be a good person to follow." I even have *babies* following me, their moms love my tweets so much!

Their customer support is fabulous; I've heard back from the company's CEO with answers to my questions on weekends!

Let's get started:

On your (preferably desktop) computer, log on to www.Social Jukebox.com.

- Next page: click on "Sign in with Twitter."
- Next page will say, "Authorize the Social Jukebox to use your account?" Click "Authorize app."
- You'll wait a few seconds while they're "Redirecting you back to the application. This may take a few moments."
- Next, it may ask for your email (or not)—but trust them!
- Next, it may say, "Your system is currently turned off. Would you like to turn it on?" Click "Yes."

Somewhere along the line, Social Jukebox, like most other programs you use, will ask if you want it to remember you on the site. Always, for services like these, check that box. It will save you time every day.

Back to the fifth bullet point above: After turning your system on, click on your image (circled below):

which will yield the screen below. Click on "Review Plan Options":

You'll want to choose the Advanced plan, circled below:

After paying, you can then go to the Jukebox Library by clicking in the circled area below:

After paying, you can then go to the Jukebox Library by clicking in the circled area below:

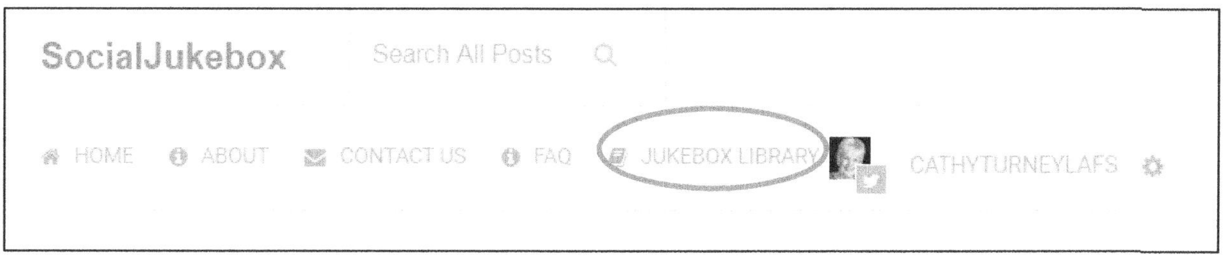

where you will be shown 18 web pages of incredible collections of brilliant quotes tailored to almost anything you could want to say. It's a gold mine! Here are a few:

Juke Box Limit Met
82 Records
BOOKSMART By Frank Sonnenberg
By: FSonnenberg
Downloaded: 50
Rating: 5.00

Brainy Quote

Juke Box Limit Met
200 Records
@BrainyQuote images
By: SEETEK_AU
Downloaded: 887
Rating: 4.83

Juke Box Limit Met
50 Records
Best of @TonyRobbins
By: SEETEK_AU
Downloaded: 261
Rating: 4.96

BRIGHT DROPS

Juke Box Limit Met
50 Records
Motivational images
By: SEETEK_AU
Downloaded: 422
Rating: 4.62

Each of the above is a "jukebox" full of quotes and images. Had you chosen Bright Drops (aka "Motivational images," circled below) it would be added to your collection of jukeboxes ("My Jukeboxes") and you'd get this page:

Click on the red-and-white jelly bean to the right of "Motivational images" above to turn that steam of tweets on. When you click there it toggles from a red-and-white to a green-and-white jellybean, and your new jukebox goes live.

Scheduling the Frequency of Your Tweets

Next, we need to schedule the frequency of your tweets. There are two ways to do this: "Auto Schedule" and "Visual Schedule." We'll do "Auto" first. In the same screen we were just working in, in the left column, circled below, click on "Auto Schedule."

Which will look like this:

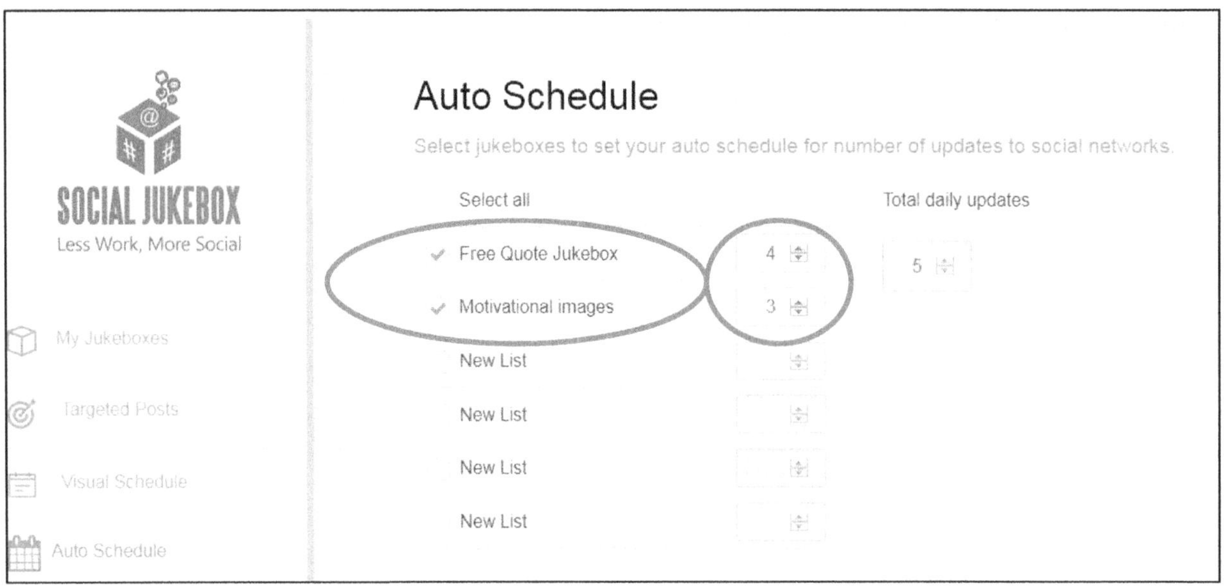

In the top part of the screen you'll see our two jukeboxes circled and to the right of them, the number of times every day that the quotes from each will be tweeted. You change the frequency of these tweets by changing the number in the boxes circled at the right. Just remember, after you finish filling the page out, click on the (onscreen) green "Save" spot.

The beauty of this system is that this schedule will repeat every week until/unless you change it. You don't have to lift a finger! It's all done for you. It's so automatic that it won't even be one of the 15 minutes a day you spend gathering your flock.

Next, let's look at "Visual Schedule."

In the left-hand column of the screen above, select "Visual Schedule."

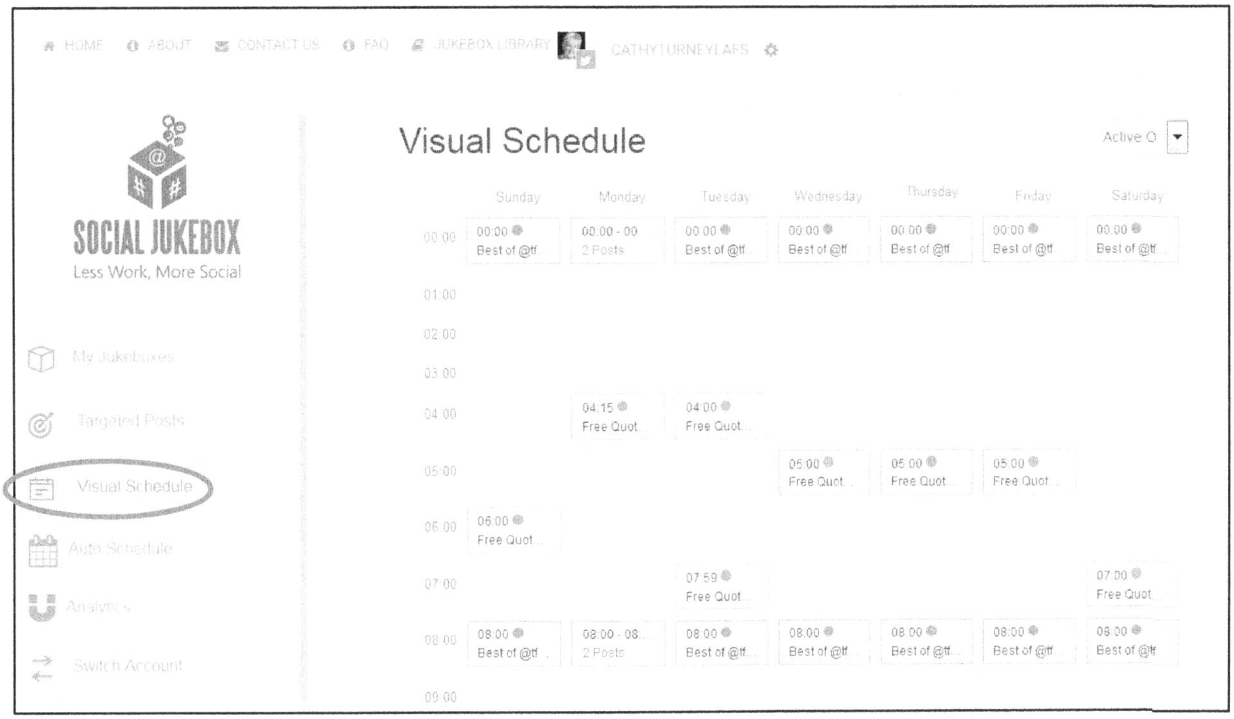

This partial screen below illustrates where you can view your whole week's scheduled tweets AND add additional tweets from your jukeboxes. For the latter purpose, I clicked in the "Sunday 1:00" area (just hover and click in that area—there is no box), which yielded this pop-up below:

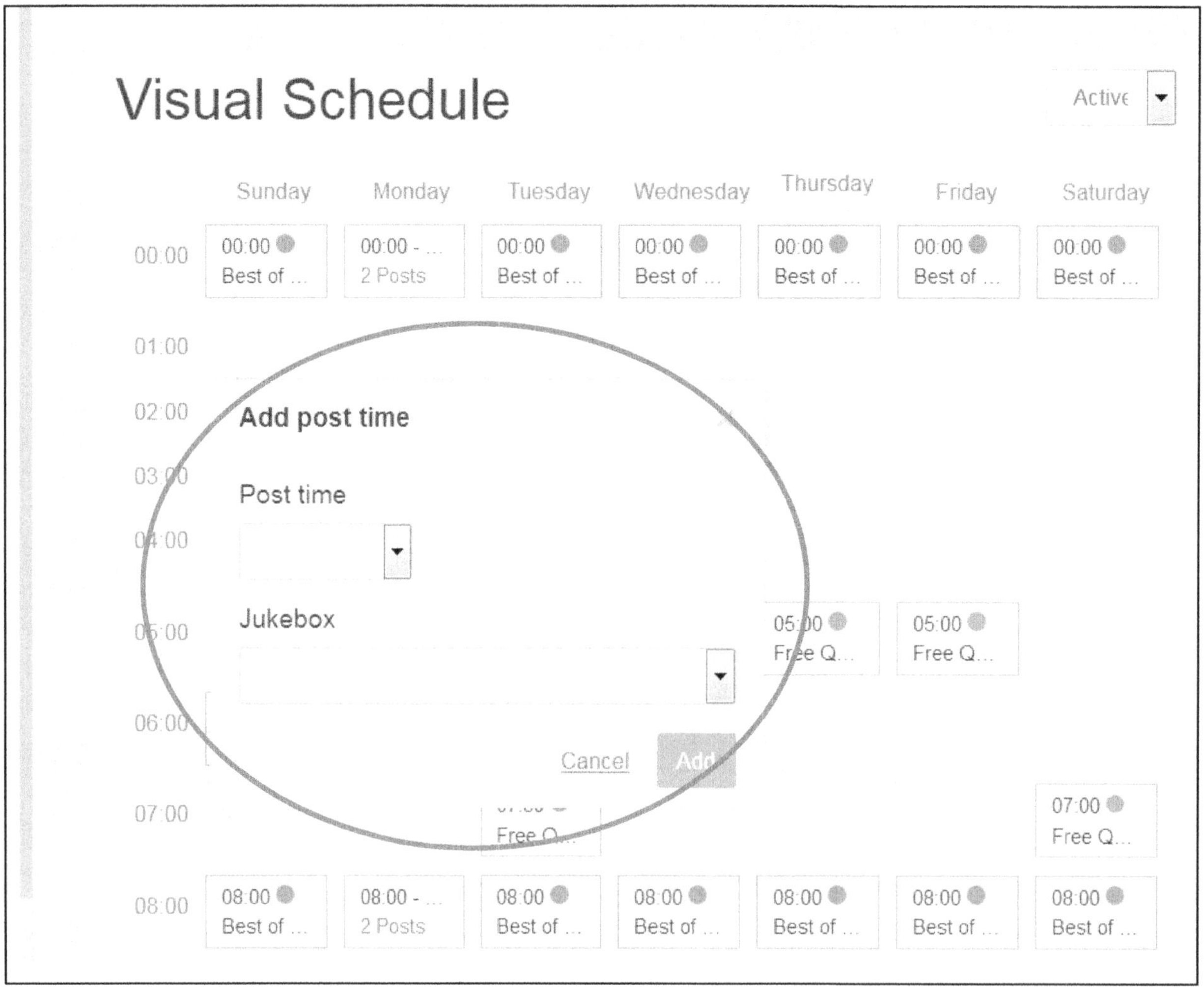

If you selected a time (click the down arrow at "Post time") and choose a jukebox (click the down arrow at "Jukebox") and then clicked "Add," an additional tweet would be inserted in the schedule:

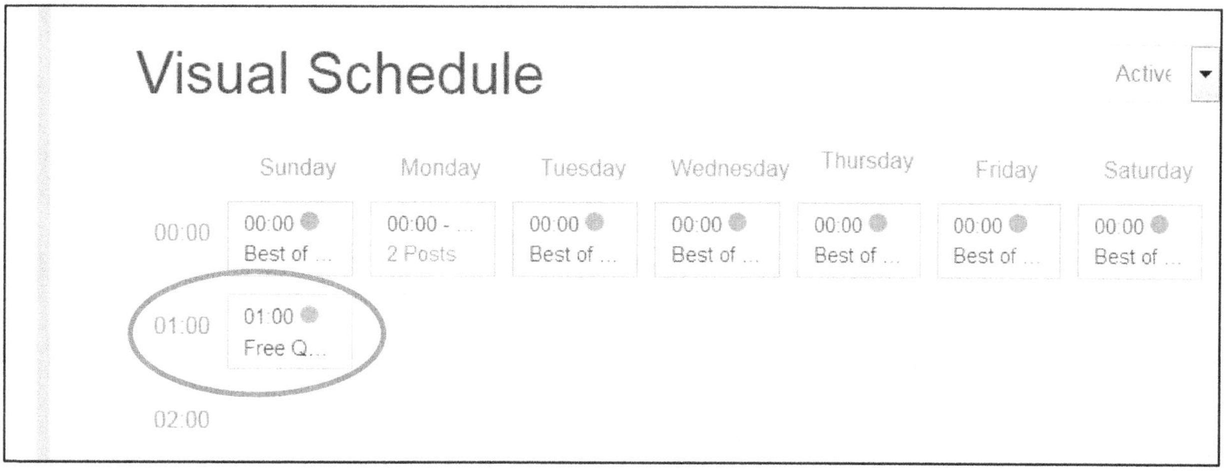

See how easy this is? Again, remember to "Add." If you're a tutorial type of person, go to YouTube and search "Social Jukebox." There's a quick run-through that gives you an overview of what the app does.

You now have glorious words to share with the world! That's all there is to hiring help to say things that make you look follow-worthy. "But what do I do next?" you ask. Not to worry—I will not abandon you, fluttering your little fledgling wings. After explaining the next program, Buffer, I will describe the 15-minute-a-day routine, step by step, which you can follow every day to get 10,000+ Twitter followers.

Now we'll log off of Social Jukebox and put it aside for a few minutes to set up our second program, Buffer.

Buffer

Buffer is where I prefer to tweet (vs. on Twitter itself) because it does additional things without any extra effort on my part, which we'll get to later in the book.

To set up your Buffer account, type https://buffer.com/ into your browser and click on "Get started now." This screen will pop up:

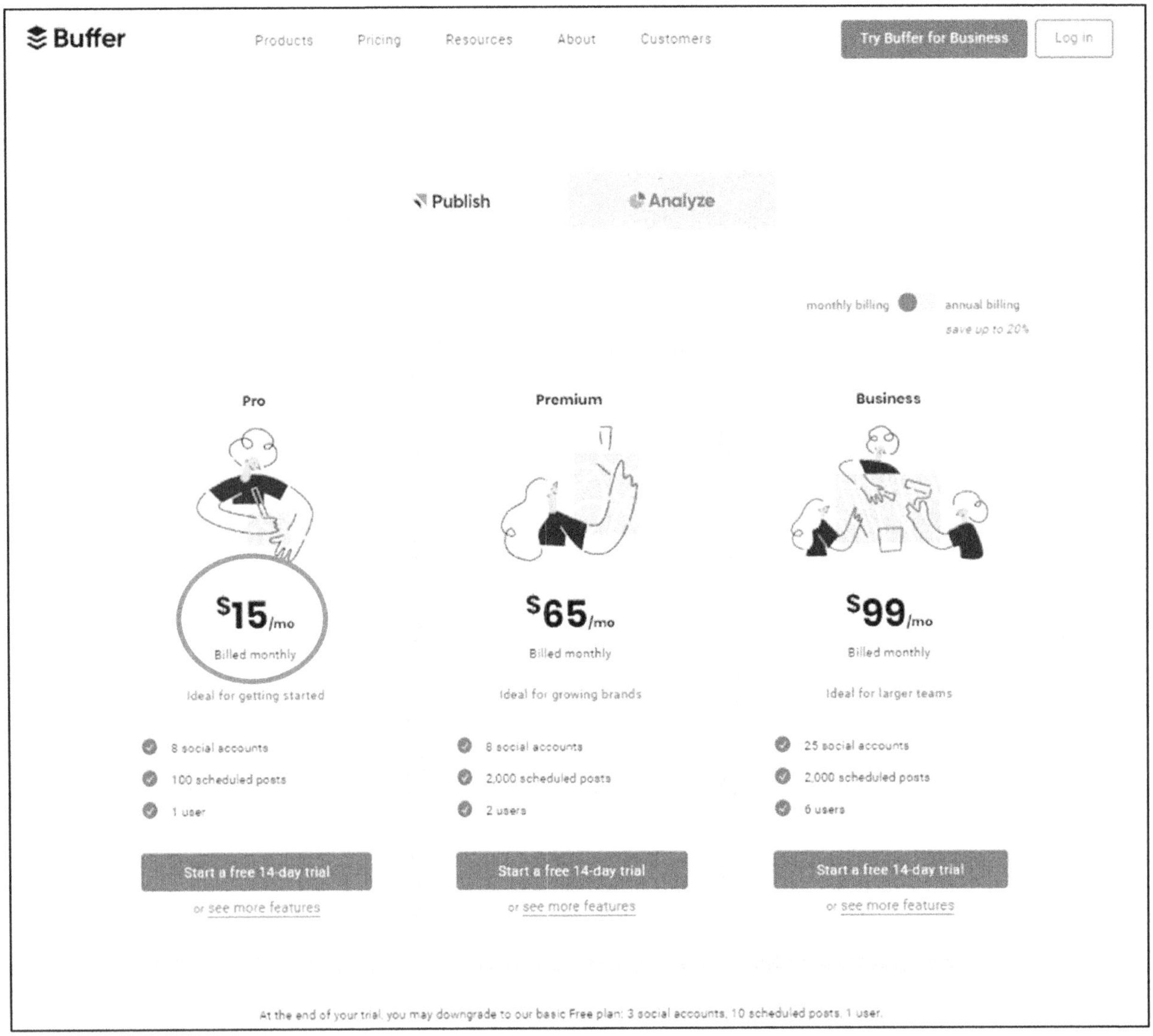

Choose the Pro ($15/month) plan which can convert to free with limited functions at the end of the trial. That's all you need. You will then be presented with this screen:

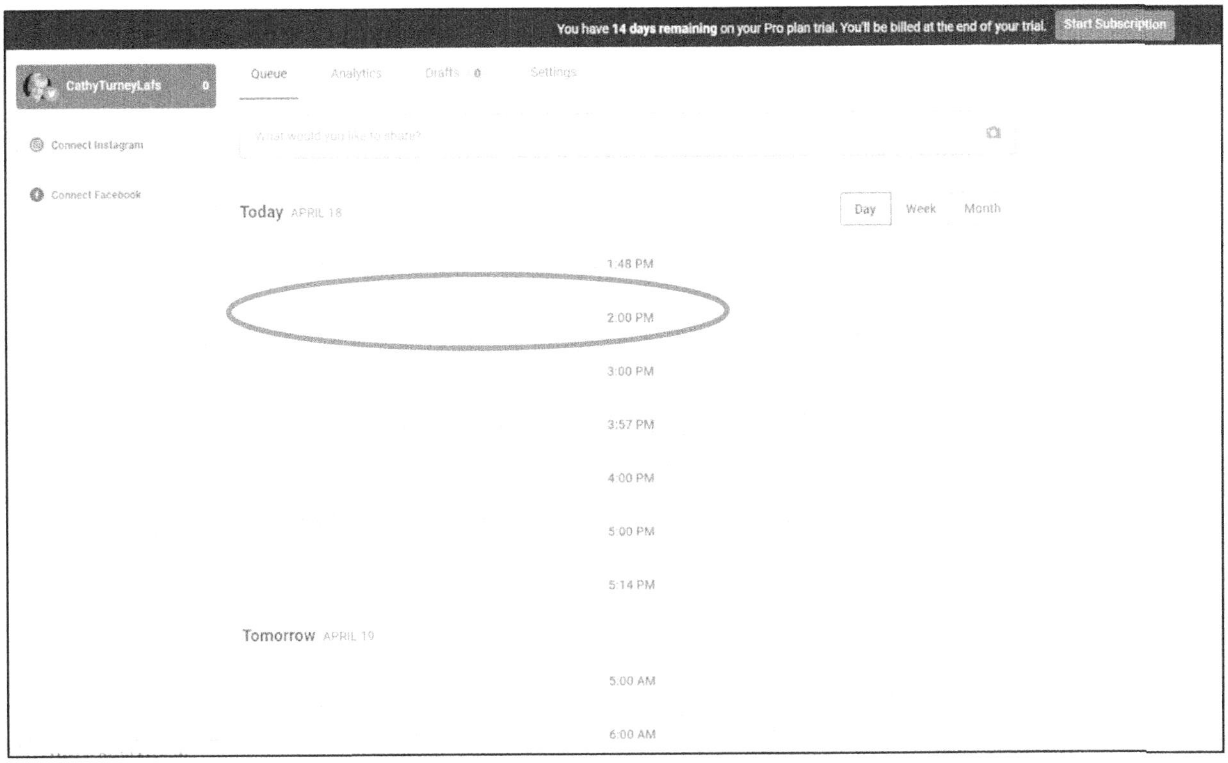

When you click on a time slot (red circled area above), the following screen will pop up:

This is where you write your tweet. As you type, the number of characters remaining will be counted down (right side, near bottom). You can change the time you want your post to go out (lower left bottom). And you can add an image. There are some great Buffer tutorials on YouTube.

The 15-Minute-a-Day System

Now I will share with you an overview of my regimen for achieving 10,000+ Twitter followers easily, quickly, and ethically. We'll get into the detailed "how to" momentarily.

Step 1: For the first step I do actually use Twitter itself. I log on and follow all my new followers back. It's faster to do it on Twitter than on the other programs.

Step 2: On Twitter, I go to My Tweets and choose the most recent tweet (automatically provided by Social Jukebox without any interference from me) that would "speak to" a large audience. The message in that tweet is what I'll base my word search on in Step 3. For example, with the Winston Churchill quote "Attitude is a little thing that makes a big difference" I would choose "attitude" as the keyword for my search.

Step 3: Next, I follow the 50 people whose tweets popped up.

Step 4: The next day I log on to Twitter to see how many followers I gained, follow them back, and repeat Steps 1 to 3.

Step 5: Clean the nest (which I do not do every day).

This is truly all I do. Every day. And once you get in the swing of it, it will only take you 15 minutes. Depending on how curious you are about what others tweet, you might stretch it out. Notice that Buffer is not included in my daily ritual—because I don't need to create tweets every day (or ever, truthfully) to amass followers. We'll get to Buffer in Part III, "Chirp It Out! Your Own Tweets."

Now let's look at these steps in detail. Remember how I kept reminding you to let the programs save your username and password ("Remember me on this computer")? If you didn't do that, add five minutes to each program so that you can hunt down your username and password, type them in, and hope your fingers typed them correctly. But it's never too late; every time you log on, the program will ask if you want to be remembered.

Step 1: Twitter—Follow new followers

Log on to Twitter. To be sure you're on the home page, click "Home," That will produce a screen similar to this:

Home
Search Twitter
Home
Explore
Notifications
Messages
Bookmarks
Lists
Profile
More
What's happening?
Marion Friedl GGultras
Trends for you
Fuddruckers
#ShadowsFX
Adam Lambert
#IfTonyStarkWereInCharge
Baskin Robbins

In the red-circled "Search" area, type "#success." That will produce a screen of people who mentioned success in their posts:

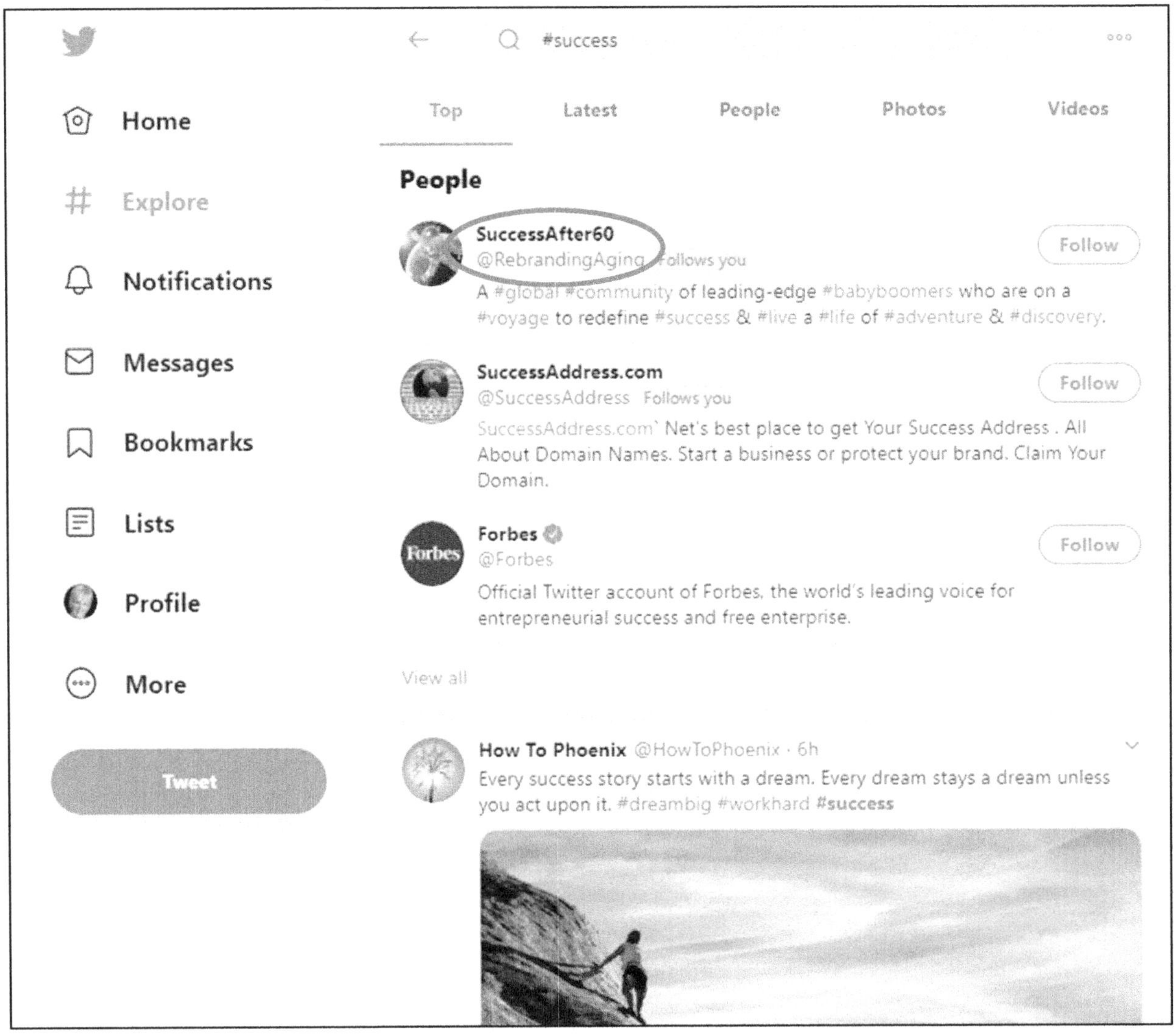

It's obvious where to follow the first three listings. But for subsequent ones, you'll need to mouse-hover over the name (circled in red). Go do the list and follow, follow, follow until you've followed about 50 people or companies. I am not selective. To do this quickly, you can't look at each one and weigh whether or not you want to become BFFs. If, while standing in line at the grocery store scrolling through tweets on my phone, I discover that I followed an ax murderer I unfollow him thusly:

Click on his handle (circled above). That will go to this screen:

And then I click on the “Following” rectangle circled, above, and it unfollows him by toggling back to “Follow,” an invitation that I would never ever consciously accept.

A little caution about follows and ratios: Rumor has it that Twitter doesn’t like you to follow more than 10 percent of your total follower count. When you’re just starting out and have fewer than 2,000 followers, you’re probably okay with following 100 each day. Regardless of how many followers you ultimately get, I wouldn’t exceed 200 follows (total) in one day. Patience! You’ll become more relevant with each passing day.

Now, more about follow limits. Remember the tortoise and the hare? Until I reached 5,000 followers, I only followed 100 people a day. Don’t be dejected—that took less than two months, and during that time I was looking more and more “relevant,” especially to others who had fewer than 1,000 followers (which is most others). Under Twitter’s Following Rules and Best Practices they have some guidelines; however, I’ve found them not to be reliable. What I’m saying is just be cautious. And if you get a warning be very, very cautious.

My advice is to be consistent in the number of people you follow and delete; increase these numbers in small increments. Don’t follow hundreds at a time after having been inactive for several days, even if it fits within the 10 percent ratio. Big shifts in numbers are like red flags to Twitter’s algorithms; they’ll send you warnings if you get out of line. If they give you too many “time-outs” they might blow your nest away, and then you’ll need to start all over again, probably with a different name, and you’ll have lost all your followers. Here is the cautious method I used:

- Up until I reached 5,000 followers I followed 100 a day
- Until I reached 6,000 followers, I followed 150 a day
- Until I reached 7,000 followers I followed 250 a day

- Until I reached 8,000 followers I followed 350 a day

Then, one day as I was merrily following 350 people, I got a major "time-out." Coincidentally, I read in the newspaper that evening that Twitter was cracking down on "aggressive" following. So I dropped back down thusly:

- 8,000 to 10,500 followers: 250/ day
- 10,500 to 11,500 followers: 350/day

Now I don't exceed 200 a day because Twitter is getting pickier about "churn," and I have achieved "relevance." Take heart and remember, your following looks huge to anyone with fewer followers!

I do repeat this routine every day, collecting followers, because it's just fun. And in Part III, we'll discuss how to make Twitter even more fun.

STEP 2: The next day I log on to Twitter to see how many followers I gained over the previous 24 hours and repeat Step 1.

Here's where you find who followed you, and you can follow them back.

Click "Notifications."

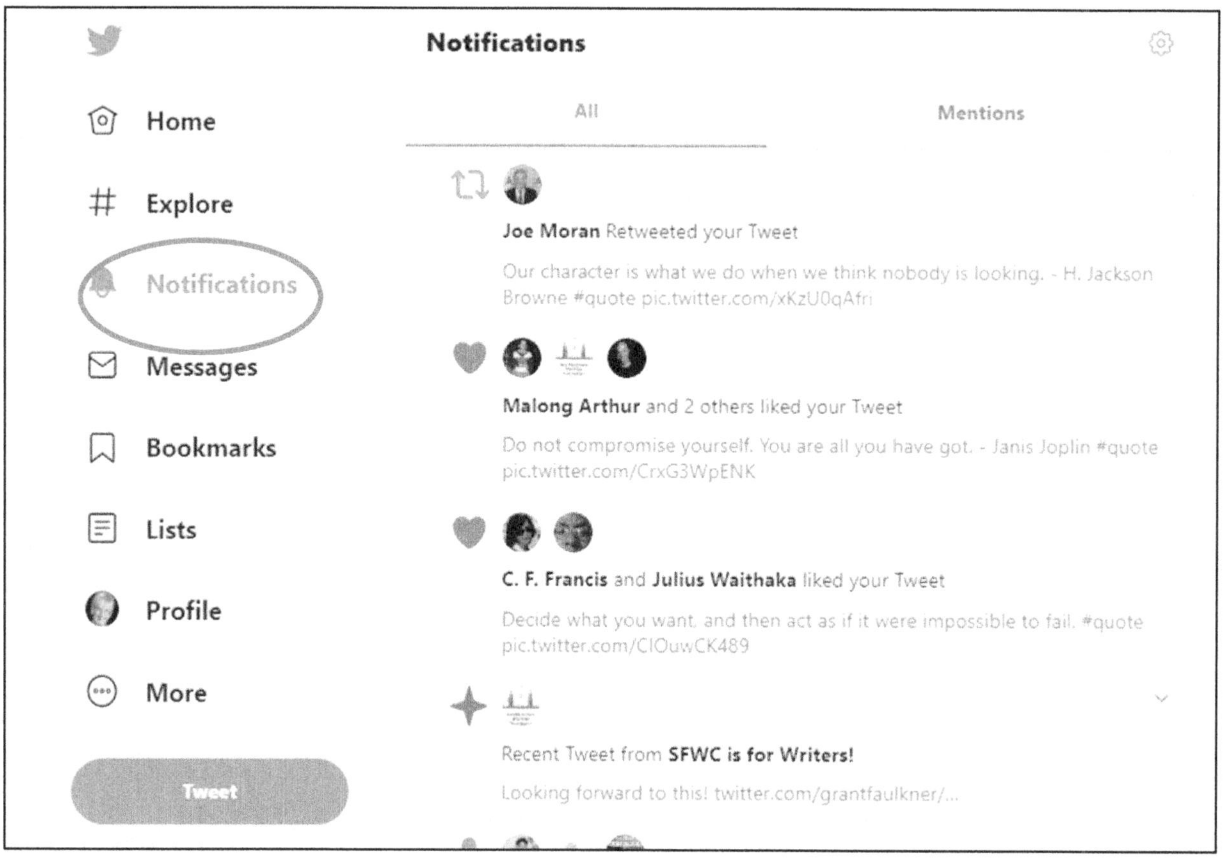

Scroll down until you find some people who followed you:

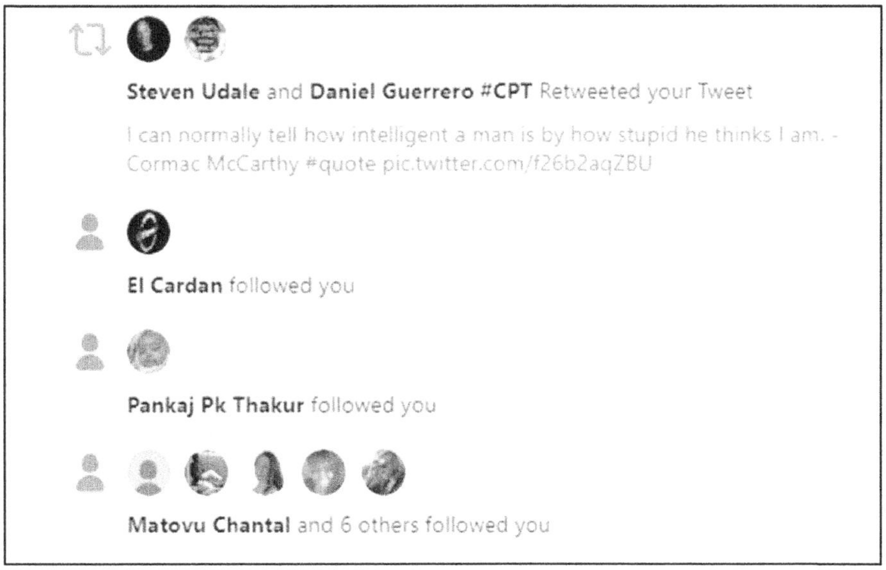

Click on their images and you'll be given the opportunity to follow them:

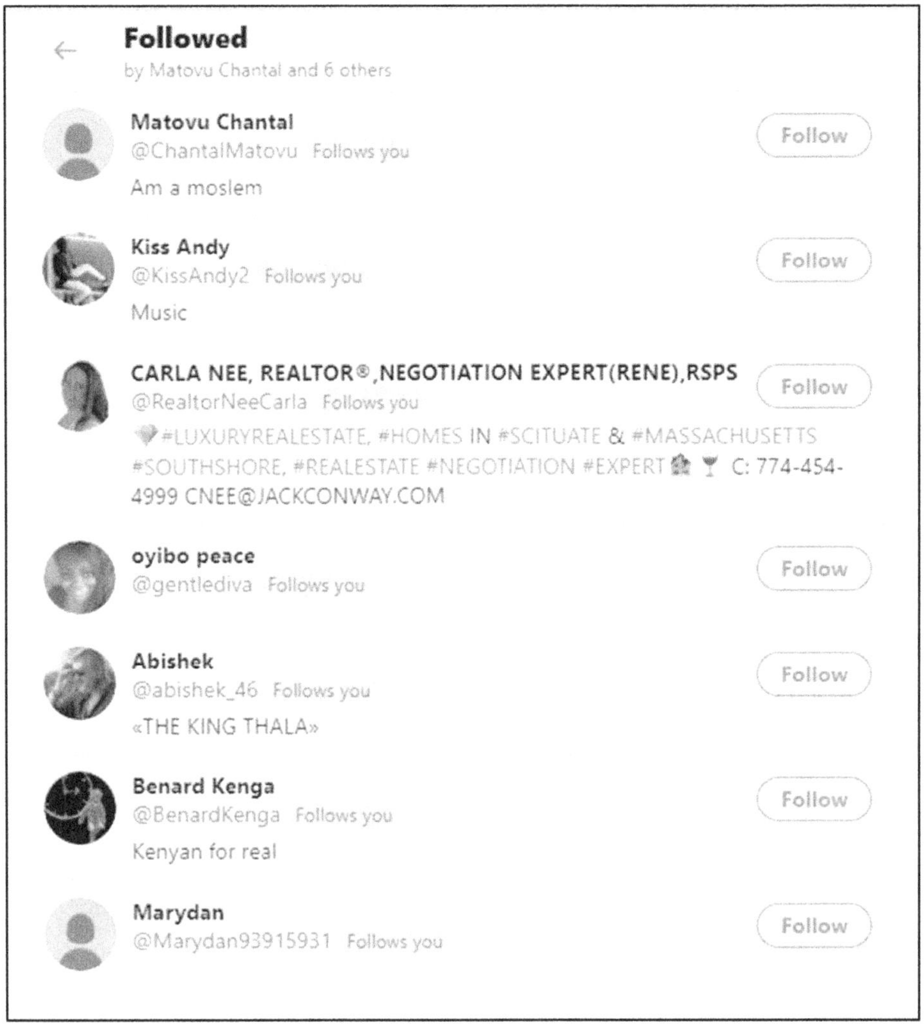

Until you hit 10,000 followers, Twitter clearly states under your avatar how many followers you have. After that, it posts 10.1K, 10.2K, etc., and you must do the mouse-hover gyration to see your exact tally. To view your actual count after 10,000, go to your home page on Twitter itself (click on "Profile at the left. Then click on your image at the right, and it will say how many followers you have (toward the bottom). Hover your mouse over "Followers", and wait a second or two; Twitter will reveal the actual number of followers you have.

Some mornings I'll log on and my account reflects 40 new followers, but when I click there to see them on the screen so I can follow them back the count is quite different. I've stopped trying to figure this out. A lot of what Twitter does defies explanation. Bottom line is how many it says I have at the top of my page in my banner—that's what the public sees.

Step 5: Cleaning the nest.

Twitter has killed all the apps that make it easy to see who's not following you back. They're trying to increase engagement at the expense of large following masses.

So what I do when my numbers get out of whack(I'm following 10% more than are following me) I just arbitrarily unfollow a few hundred.

I know that sounds heartless, but if you're after follower count, you might need to do this. Here's how: In that left column that we keep going back to, click on "Profile." You'll wind up here:

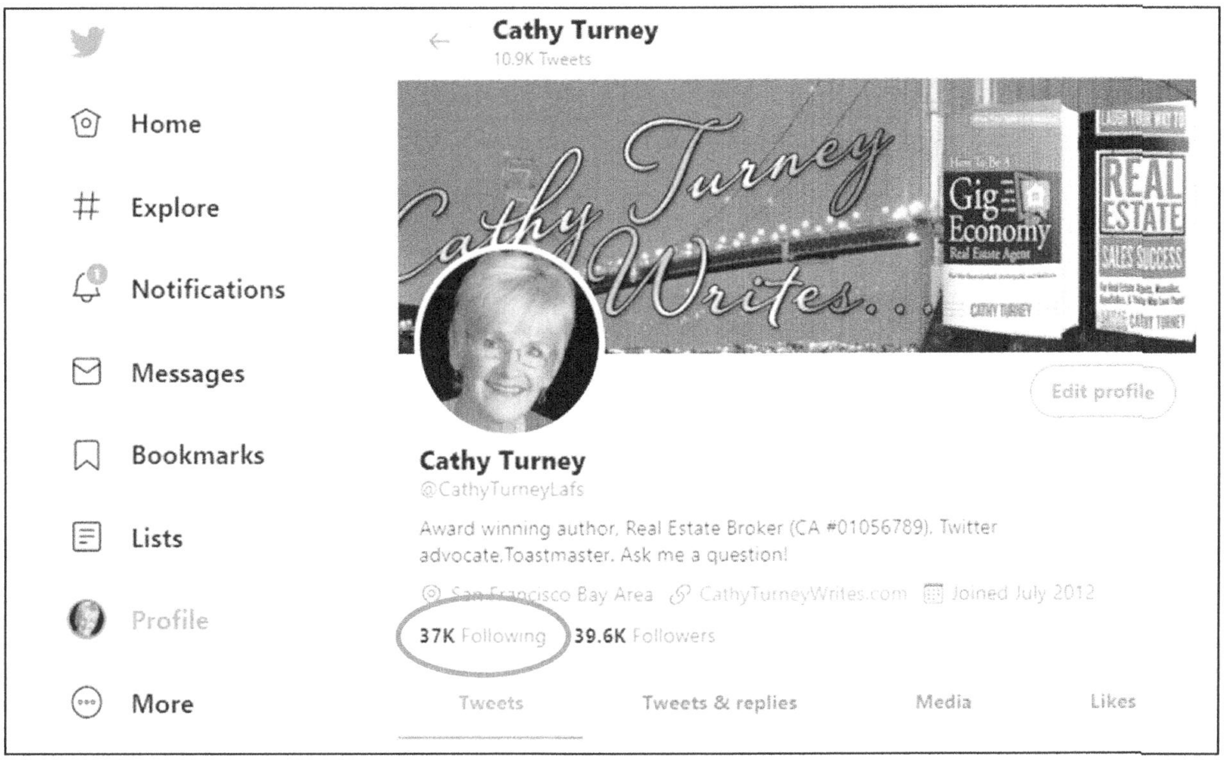

Click on "Following" (circled in red above). That will produce this:

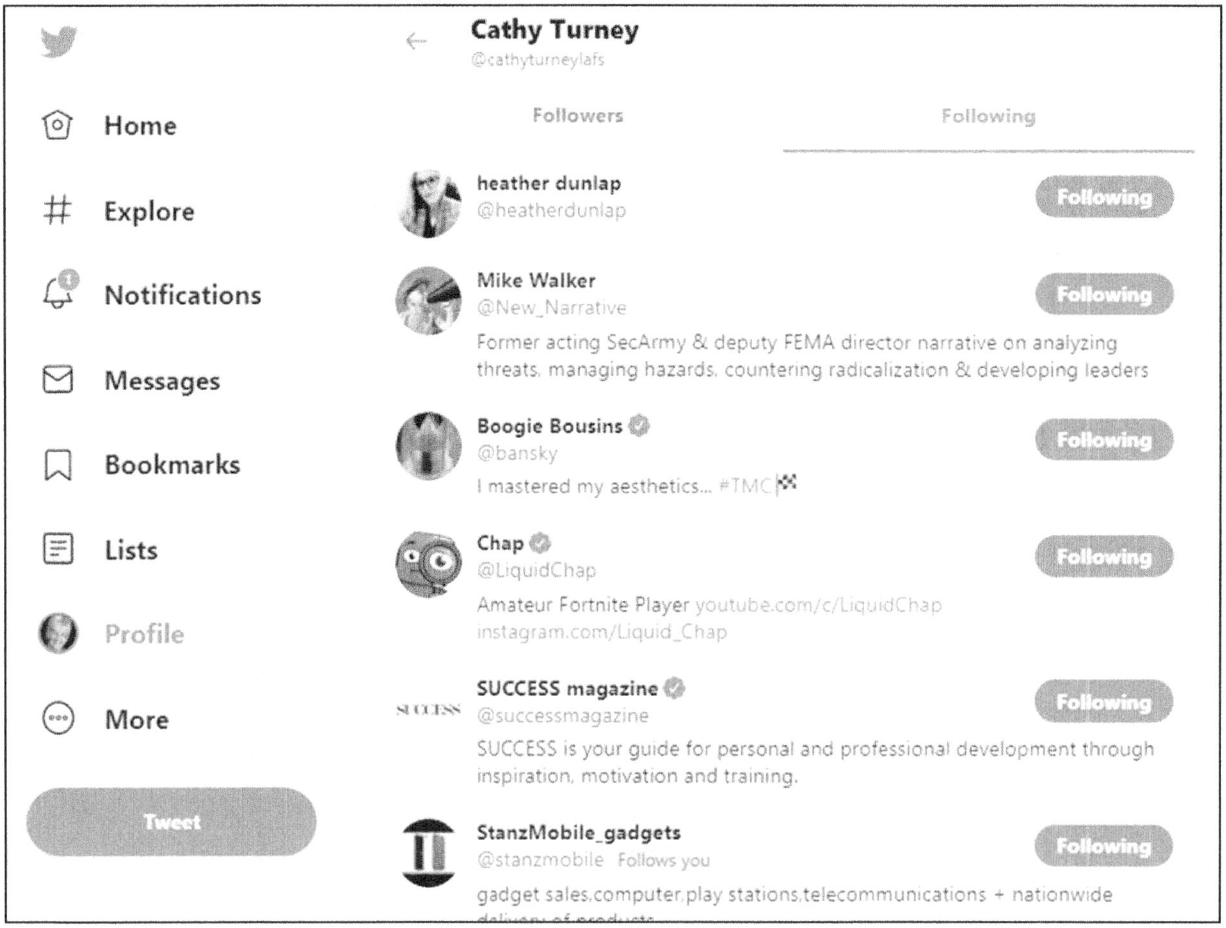

Then just click, click, click "Following" and it will toggle back to Unfollow.

If there are people that you want to continue hearing from, don't unfollow them.

You have now completed all the steps necessary to gather 10,000+ Twitter followers easily, quickly, and ethically. I know you'll have great success. It took me less than a year (during which I was looking more relevant each day) to accomplish this goal, and I know you can, too. But you must do it every day (time off occasionally for good behavior). It builds exponentially, and if you don't keep feeding and culling the flock your numbers will actually decline. After a while, when you get your routine down, you'll look forward to sitting at your desk every morning and doing it. Really!

In Part III we'll build on what you've learned here and soar. I recommend that you get set up and become comfortable with Parts I and II before distracting yourself with all the fun in Part III. It's not that any of it is difficult; it's that you want to remain focused on getting (and keeping) 10,000+ Twitter followers easily, quickly, and ethically…until you could do it in your sleep.

PART III – Soaring

"Hawks will soar above the morning lark."

—William Shakespeare, *The Taming of the Shrew*

You Are a Hawk!

Congratulations on positioning yourself to become relevant and to soar. But don't just take my word for what Twitter can do for you:

If your company is not using Twitter then your company is going to ~~miss out~~ be crushed.
—Aaron Lee (social media expert, entrepreneur, writes for *Forbes, Inc., Huffington Post*)

More of my business comes from Twitter than from any other platform.
—Frances Caballo, social media strategist, manager, and author of *Social Media Just for Writers*

If you're on Twitter, what you're saying is, "I'm important enough for you to care what I think."
—Donald Glover

We live in a social world now, and there's no denying the power that Twitter has yielded across all verticals. Sports is a perfect fit because fans are highly emotionally charged and things happen quick.
—Amy Jo Martin, ranked third most powerful woman on Twitter by *Forbes;* author of *Renegades Write the Rules*

Twitter users are more likely than Facebook users to shop online. If you want access to the online shopping crowd research from GlobalWebIndex found that 73% of Twitter users shop online each month (compared to 69% of Facebook users).
—Jason DeMers, CEO AudienceBloom

My (Twitter) followers have helped me grab the attention of powerful influencers, got one of my stories 30,000 reads on the story-sharing site WattPad, and landed me a job at a publishing house.
—Daniel Parson (@DKParsonsWriter) via Jane Friedman's Sept. 6, 2015, post "Using Twitter to Make Powerful Connections As a Creative Professional."

Twitter is THE place to be if you have *anything* to promote or sell.

I'm in Love! The Engagement

By now, can't you envision a future with Twitter? Meeting birds of a feather, flocking to events you hear about in your stream, ruling the roost as a Twitterati?! Heretofore, you did not need to get "engaged." And if your sole reason for being (on Twitter) is to be big like a condor and inspire awe, you still don't have to get engaged. Because as long as you and/or Social Jukebox are posting, following, and unfollowing regularly, your brood will grow and grow.

But this one-sided relationship may not lead to happily ever after if you start to feel there's something missing. I mean, *where's the love?* That's when you'll want to get engaged.

Notifications and Messages

These two spots on Twitter are where people communicate with you. On a desk-top or laptop, they're at the left of a screen. On the iPhone it displays across the bottom of the screen regardless of whose page you're on.

When you click on Notifications, a list like this will pop up:

As you see, Notifications are where Twitter tells you who likes and retweets your posts. It's also where a tweet that includes your hashtag (i.e., @CathyTurneyLafs) will appear. It means that someone is trying to communicate with you directly. Most of those direct tweets will be wild geese just trying to get some interaction going. But occasionally, you'll receive a notification that you would want to act on, such as this one from an author friend of mine:

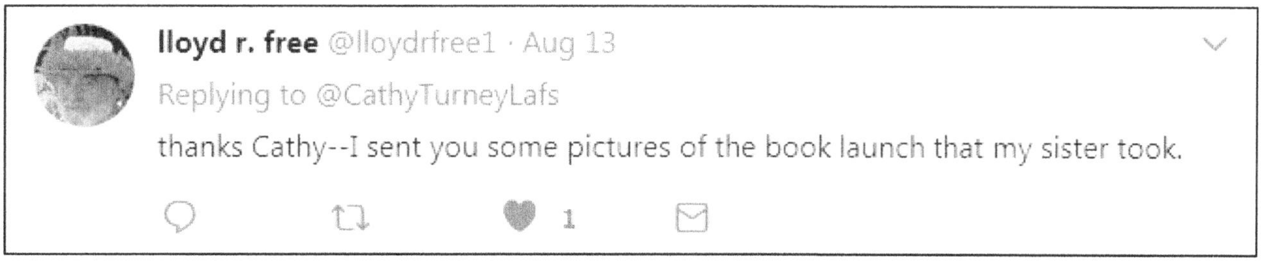

So it's good to check. Notifications are the equivalent of a Facebook comment; anyone can see them.

"Messages" are private (the Facebook equivalent of a direct message), and display thusly:

Strangely, the communiqués you get in messages are *not* that personal, unless you count marriage proposals and solicitations for bail money. I'm often asked, what is proper Twitter etiquette in responding to questionable tweets. Do not feel that you must respond. They aren't expecting you to. You want the kind of engagement that leads to happily ever after. If it makes you feel better about ignoring them, I rarely reciprocated until I reached 10,000 followers.

Chirp It Out! Your Own Tweets

Social Jukebox is a lifesaver and will, all by itself, get you to—and above—10,000 followers easily, quickly, and ethically. But equally important (or why would you bother trying to look relevant?) is getting the word out about what you do or sell.

Although the process of amassing your flock is more easily accomplished on a laptop or desktop computer, sending out a tweet from your handheld device is simple. Just log on to Twitter and look for this little square with the quill pen in it , click, write 280 characters or less, then tap "Tweet" in the lower right corner of the screen. Twitter will count down how many characters you have remaining and won't launch your tweet if it exceeds 280 characters. You'll figure this out very quickly.

My preferred method of tweeting my own messages, however, is on my laptop using Buffer because I can easily add graphics and links on their platform, and Buffer will shrink the links to minimize the character count (instructions below). For the venturesome, this is where you can also schedule tweets to appear at a later date.

Buffer can be a little tricky, but here's how to deal with it. To post your own tweets on Buffer log on to their site, and you'll land here:

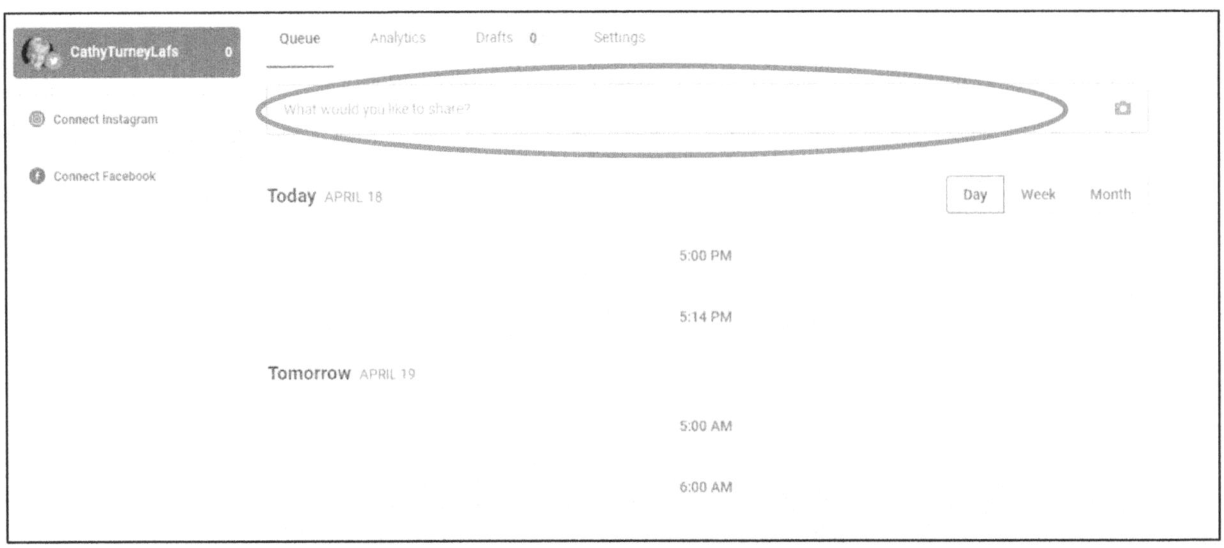

Buffer allows you to post to several social media sites at once, as long as you select them. For this exercise, we'll only post to Twitter. In the above screenshot, click in the red circled area. That will yield:

When you type in the text area above, the light blue jellybean will turn to bright blue. Hitting the down arrow will produce this:

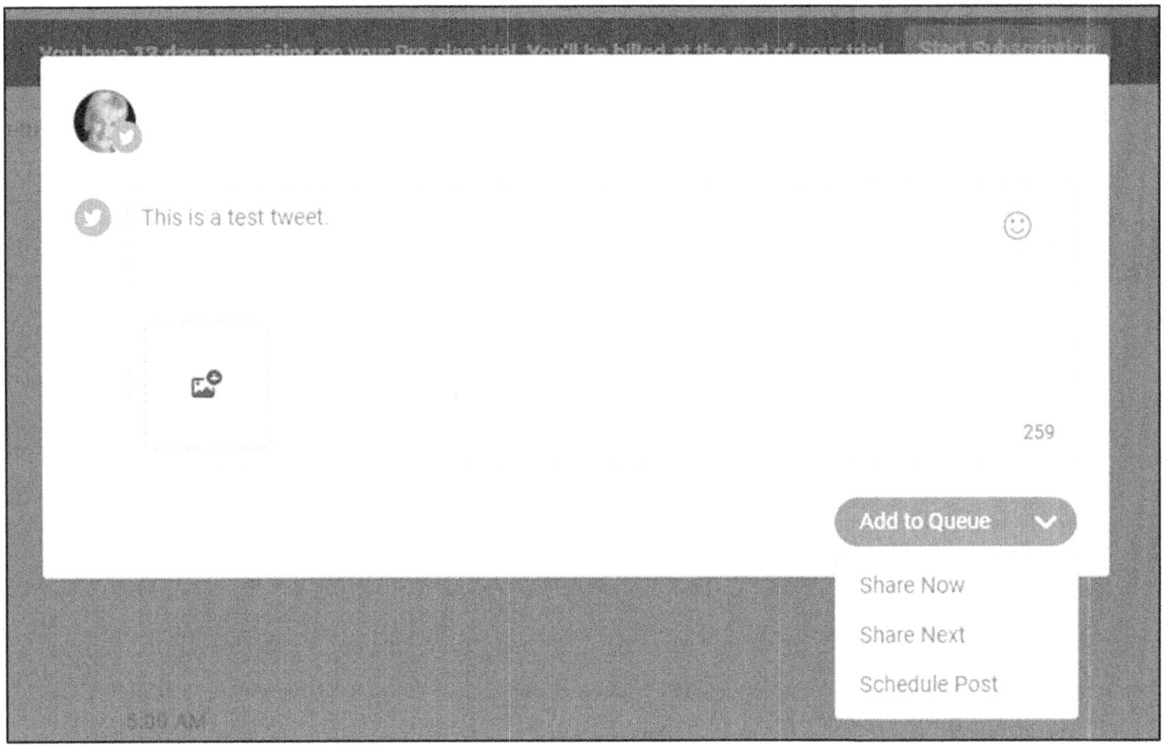

- "Share Now" will post the tweet to Twitter immediately.
- "Share Next" will put it in the schedule you can set up.
- And "Schedule Post" will land you on a calendar where you can choose the date to post.

A little obtuse, but keep in mind that you really don't need to use Buffer to get to 10,000+ Twitter followers easily, quickly, and ethically. I like it because I can write a bunch of tweets and have them go out at a later date. If you're interested in that function of Buffer, do watch some YouTube videos (go to www.YouTube.com and search "Buffer.") I would only confuse you.

An especially nice thing about Buffer is that if you want to add a web link to your tweet, simply copy its address from your browser and paste it into the tweet; Buffer automatically shrinks it.

Besides hunting for, saving, and uploading images of your own, there is another nifty way to increase the visual appeal of your tweets: the GIF button. It's easy to use when posting from Twitter itself. The GIF (stands for graphic interchange format) button works on all devices: laptop, desktop, handheld. Let's practice GIFs and sending tweets from a handheld device:

Log on to Twitter. Regardless of where that lands you, along the bottom of the screen will be some icons and you want to choose and click "Home" in the left corner.

Next, click on the quill pen box in the upper right-hand corner of the screen, which will yield the "What's happening?" or "Compose new tweet" screen, shown below:

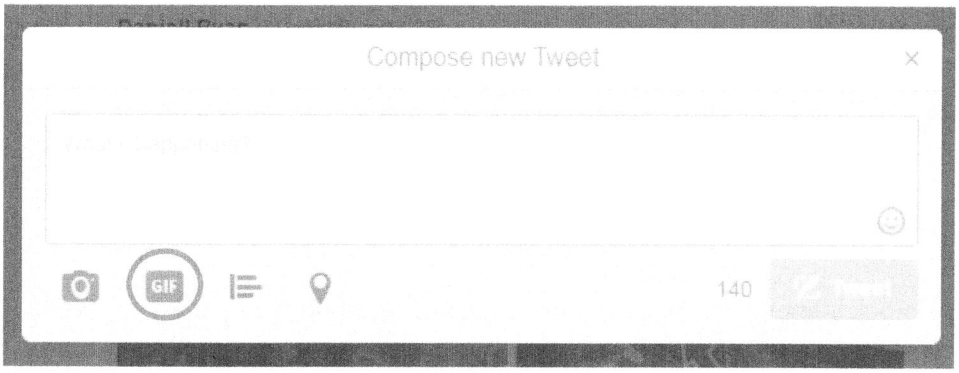

That's where you write your message. At the bottom of that screen is the GIF button (circled above). Clicking on that will produce:

Clicking on one of the categories will produce lots of neat images for you to choose from, many of them animated. Choose one by tapping it (sometimes you have to tap it several times before it works…technology ☺). Then, finish writing your message and click "Tweet." It's very fun.

That's really all there is to creating your own tweets. If you don't see what you just posted in your "Tweets," click on "Tweets and Replies." It's there somewhere!

But, what do you *say* when you write your own posts? In the interest of KISS and staying on purpose, I'll not spend a lot of time elaborating on that—there are lots of *othe*r books about it. But, here are a few pointers:

First, always observe the 80/20 Rule: promote yourself 20 percent of the time, and talk about anything else the rest of the time. People don't like to be sold to.

Subjects to tweet about can include announcements, happenings in your friends' lives and your own, birthdays, news events…the list goes on. Look at what others post. There's no end to subject matter. Just don't overdo the gratuitous stuff. Twitter is so past the what I ate for breakfast fodder.

Times and Days of the Week to Tweet

True, it's always five o'clock somewhere, but you need to concentrate on the time zones where you want your message to be heard. I recommend 6 a.m., 9 a.m., noon, 3 p.m., and 5 p.m. in your own time zone. If your product has international appeal, target other countries' time zones occasionally—or extend your daily tweet schedule to include daylight hours in those countries. Regardless of when you post, your tweets will appear in your followers' streams, but you want to be as close to the top, chronologically, as possible.

Additional Ways to Increase Your Followers

As explained in Part II, searching for people to follow by using standout words in your latest tweet (which Social Jukebox posts for you) is my modus operandi; I do this every day. But here's an enhancement to that:

When a major event occurs and I know a lot of people will be tweeting about it, I'll do a search for that topic. In response to one such incident in the news over the summer, I searched #hope and found 690,000 tweets with that same hashtag over the previous six hours—pretty impressive numbers. I followed 200 of those tweeters and gained 62 followers that day.

Another way to add followers that doesn't eat up your ratio:

1. On Twitter, click on "Home."
2. Search a topic (i.e., #nonfiction).
3. Scroll down, looking for tweets that have few (or no) likes.
4. "Like" them, because those tweeters will notice every new tweet and probably follow you back.

The reason I recommend doing this from the feed that results from a hashtag search vs. your home stream is that the home stream consists of tweets from people whom you already follow and who are likely following you back. On the hashtag search feed you'll run into both, but the nonfollowers will outnumber the followers you already have.

Pin That Tweet

You will get to 10,000+ Twitter followers easily, quickly, and ethically if you never ever ~~stick a voodoo pin in a nonfollower~~ pin a tweet. But doing so will boost your numbers, and it's easy!

Pinning a tweet means that you stick it permanently (until you unstick it or replace it with another) at the top of your tweet stream. When potential followers check you out, they'll see that warm and feathery tweet at the top of your page and know that you are a wonderful human being, worthy of following. Here's how to do it:

Click on "Profile" in the left column on any Twitter page. This will land you in your tweet stream. Here are a couple of tweets from my stream:

Clicking on the down arrows (circled above) will produce this:

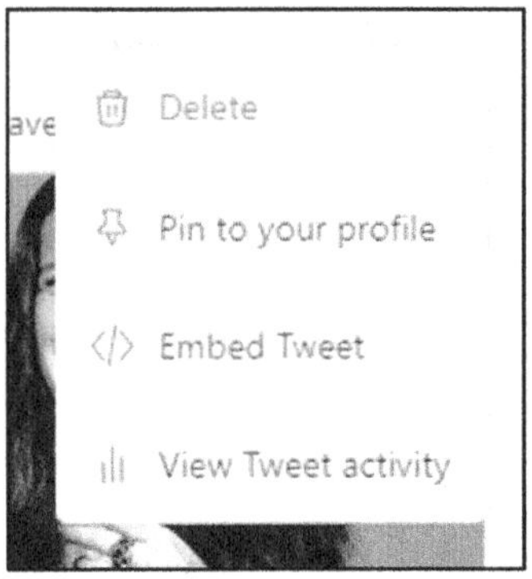

That is all there is to it. When you want to replace a pinned tweet with another, simply repeat the process above; you do not need to unpin the previously pinned tweet. If you want to unpin a tweet and *not* replace it, clicking on the down arrow of the pinned tweet will give you the option to unpin it.

Gather Your Flock—Lists

As you can imagine when you get thousands of followers who post an average of three times a day, your stream will become a raging river. And you'll notice that some of your followers say things you don't want to miss. Well, now that you're relevant, you can take some time to play in the river—to choose a bunch of those tweeters to put into your own private birdbath—by way of "lists."

I am going to make lists easy for you and fun, even, because I spent ~~thousands of~~ hours interpreting Twitter's obtuse instructions about how to make and use them. O.M.G. If Twitter could just speak right-brain English. But it can't.

To create a list, log on to Twitter and click on ""Lists" in the column at the left. That will present a screen similar to this:

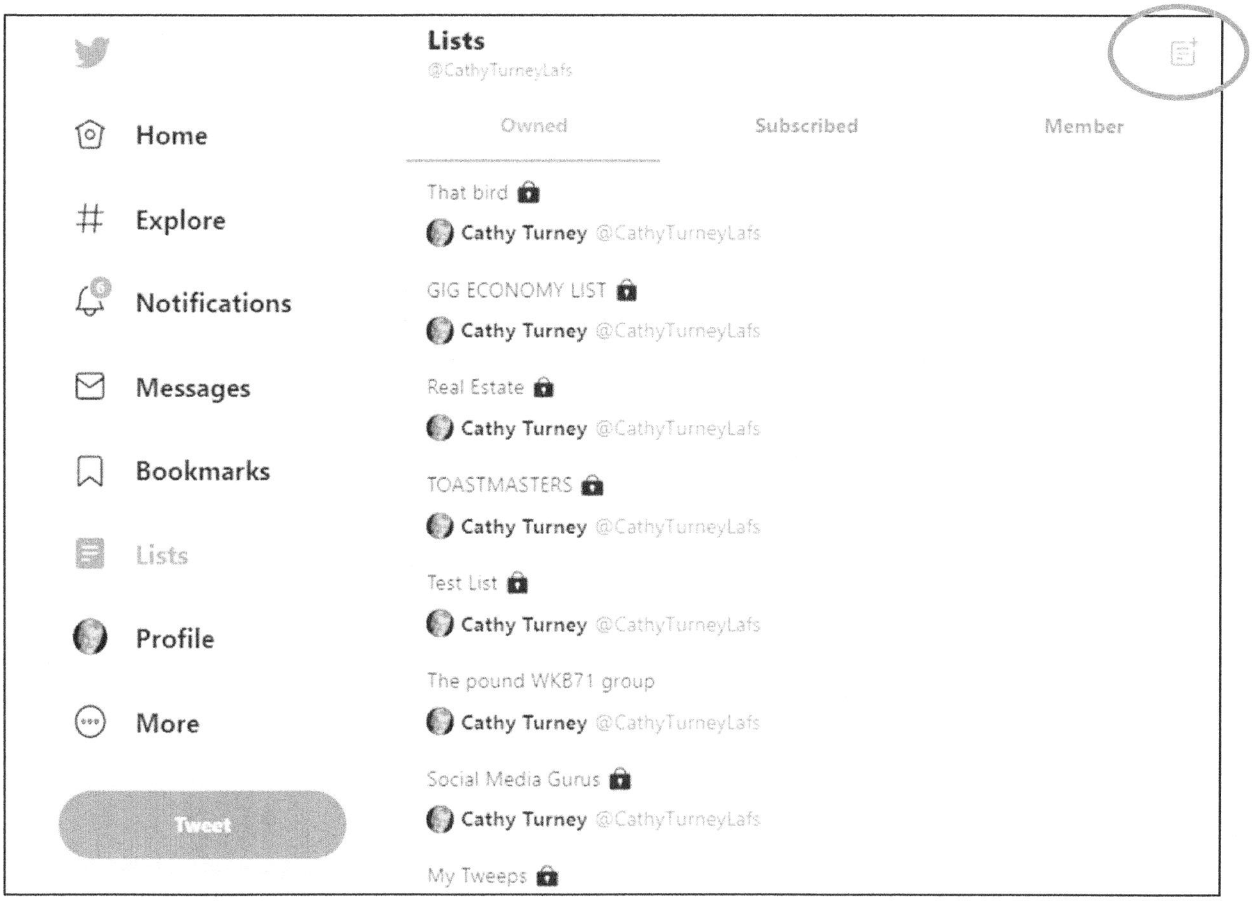

Now, click on the little icon circled in red above. That will produce:

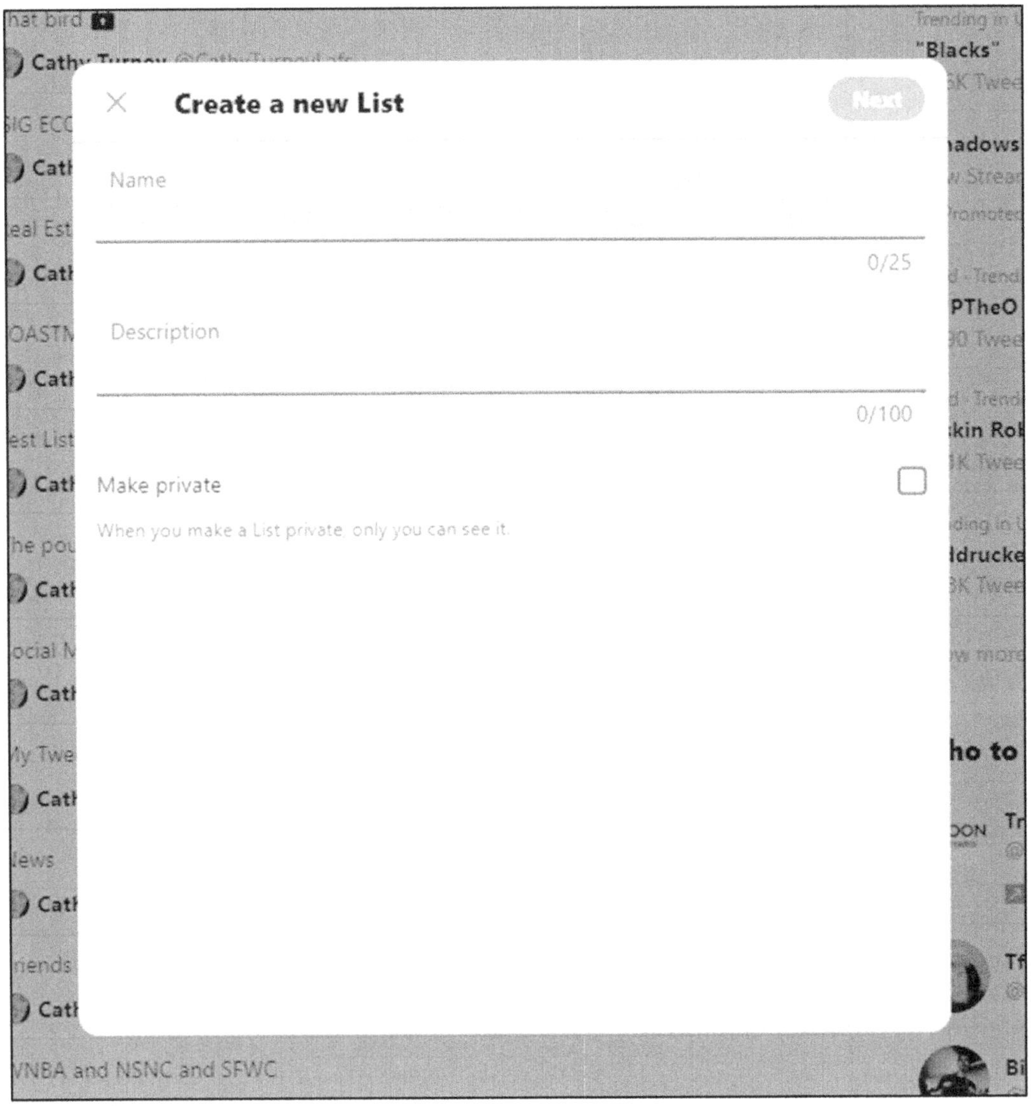

Create a name to call your list and fill out the information above but DO NOT exceed 25 characters for the list name or begin it with a numeral. Choose “public” if you want the world to see whom you follow or “private” if you don’t. “Next” will produce a list of all your followers and you can select which ones you want to add to the list:

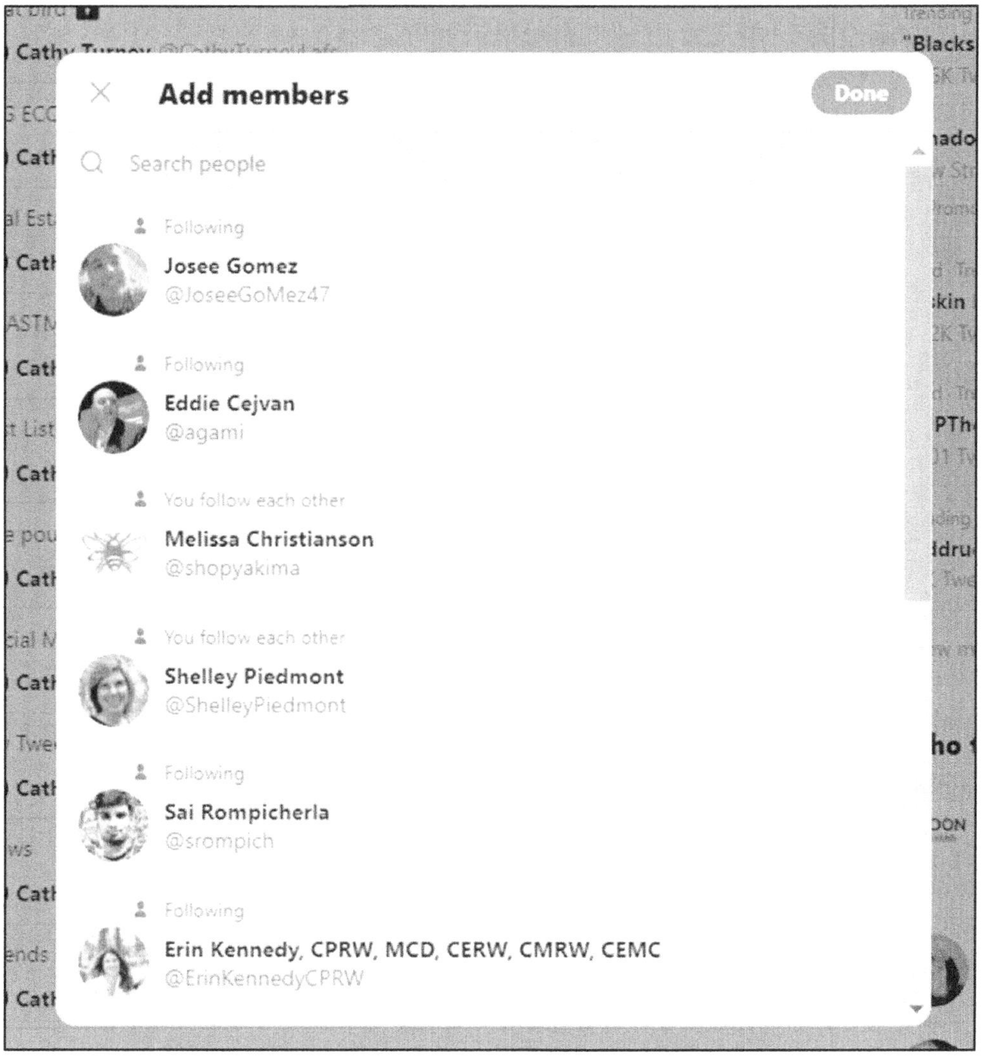

But there's another way—in case you want to go back to a list that you previously created and add people to it.

Click on "Home" in the upper left corner of your screen, which will produce this:

In the top line's search area (circled) type in the name of the person you'd like to add to a list. I want to be sure I stay on top of everything Lisa Tener has to say, so typing in her name yielded this screen:

Click on the three horizontal dots under her banner. This drop-down menu appears:

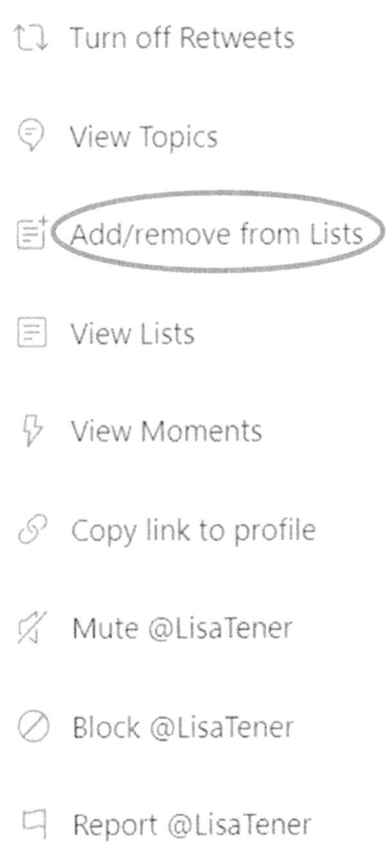

Click "Add or remove from lists" (above), which yields the list you've already created in your own lists, like mine below:

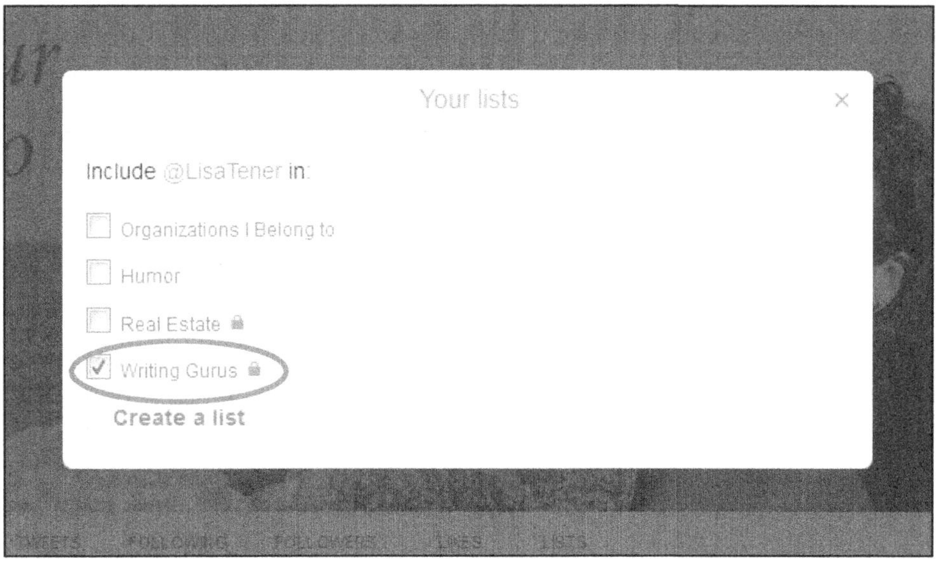

I wanted to add her to my Writing Gurus list, so I checked that box and "X"ed out of the screen. Now I know I won't miss a brilliant pronouncement that Lisa tweets.

To delete a member on a list, simply repeat the above steps (home, search name, click the three horizontal dots, click "Add or remove from list") and uncheck the box (it's a toggle switch).

The key thing to remember is that when you are creating a new list you must do that from *your own* page. When you add to or remove someone from a list, you do that on *their* page.

Now! Where do you find those lists you created so that you can check your gurus' posts? First, we'll do this on the laptop (easier to follow), and then I'll describe how to do it on an iPhone. I apologize that I can't instruct you on any other handheld device, but if you Google "How to view lists on a Samsung" (or your device's brand) you should find instructions. It's great to be able to pull up your favorite tweeters on your handheld device because you can check in on them while you're out and about. On the laptop, when you log on to Twitter, in the left column, click on "Lists." All the lists you've created will display, and you can just choose whichever one you want to view.

On your handheld device, follow these steps:

1. Log on to Twitter
2. Click on the "Home" icon at the bottom of your screen
3. At the top of the next screen. Click on your image.
4. A pop-up menu will appear, and choose "View Lists"
5. There they are!

Heads-up! You may be surprised to see tweets by people you have not put on that particular list. That happens when the person you did include does some retweets; all those retweets will be included in their (thus, your) stream. I wish Twitter would change this because some folks are serial retweeters and that seriously mucks up the stream. Ostensibly, all the retweets are things you would be interested in by virtue of your having included the retweeter on your list.

Lists are a wonderful thing.

Hashtags

As you read in Part I, hashtags pull us into a conversation about whatever word or topic follows that hashtag sign (e.g., #baseball will take us to a stream of tweets that mention baseball). Hashtags are your ticket to targeting information and news you want and finding more soul mates to chirp with.

Say you want to hear a discussion and meet some people involved with cooking. Log on to Twitter, go to your home page (below) and in the upper-right-corner search area circled, type #cooking. A list of the most active hashtags and tweeters involved with cooking will drop down:

The top four listings, which begin with a hashtag, are conversations that have the word *cooking* in them:

#Cooking

#cookinggap

#CookingFever

#cookinginthebarrio east los ponygirl

The next six listings, which begin with an @, indicate people/companies/groups who tweet about food:

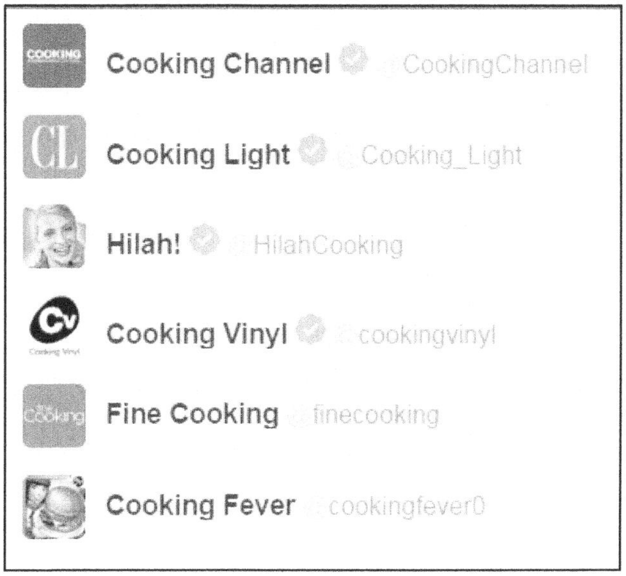

FYI, the little blue check mark indicates that Twitter has verified that that name belongs to the tweeter using it. Don't be too concerned if you don't see a blue check mark; it's really to distinguish public figures from impostors.

Clicking on any of these will land you in a stream of tweets all about food.

Easy. Fun. Too fun…set a timer.

Before you hashtag and rehash, define your purpose in searching. If you're a writer, #amwriting may not be the most productive place to find readers and buyers for your books. You'll meet lots of other people there who are doing what you do—writing. I love #amwriting and enjoy its camaraderie, but to promote my real estate writing I focus on streams where my readers might go: #realestate, #investing, #homes, and others.

If you have a product or service to sell, the main benefit to your being on Twitter will be looking relevant enough to make people want what you produce. And you do that by amassing followers and occasionally posting tweets that you create yourself (as opposed to the Social Jukebox quotes), like testimonials from happy clients, awards your wine won, public service your company does. If you have a landscape business you might search for hashtags like #gardening, #landscapedesign, but remember that the most important thing is that you are posting consistently, not necessarily the hashtags you pick to put in your tweets.

"Eight Days a Week" ~~by the Beatles~~

There have been short blocks of time when I did not follow anyone. The momentum carried itself for a few days (thanks to Social Jukebox), but when I disappeared for two weeks during summer vacation sans Internet connection, I dropped from 6,407 followers to 6,395. Not horrendous, but it shows the importance of performing the ritual every day.

Contrary to what some Twitter gurus say, I've found that following on weekends can yield results. I think you get non-business people who are just trawling for fun. If your banner looks engaging, and you follow them, many will follow you back. (But, remember—don't mega-follow on the same day that you clean house.)

Don't be paranoid about not always returning favors when others tweet to you ("follow me on Facebook" or "download my amazing book," or thanking people for following you or for their retweets, etc.). We're striving for speed here. Again, it's all about the bottom-line numbers. The time spent guilt-tripping could be spent following others. And it will certainly exceed 15 minutes a day. Seriously, until you become "relevant" you must focus.

Let My Numbers Convince You

- One Tuesday I searched #Tuesday, followed 150 of those tweeters and netted 84. Way more than I got with #MondayMotivation (30); there's just something about Mondays.
- On Black Friday I searched #BlackFriday and tweeted about my real estate book to 250 people. I added 155 followers! It had some residual effect, too, because the next day I was up another 48 followers. Don't go crazy with the self-promotion, though. Remember the 80/20 rule.
- I am a dog person, but I suspect that there are more cat lovers on social media. So I retweeted a cat video. I did. It netted me 80 new followers.
- On New Year's Day I searched #stress and followed 400 people and netted 75 new followers.

I say "net" because God only knows how Twitter computes your bottom line. I may log on and see that yesterday I had 10,323 followers and today I have 10,373 followers (up 50), but when I go to follow them back, Twitter only shows me 35 new followers. One day Twitter said I had

10,516 followers, then next day 10,496 (down 20), but when I clicked on "new followers" there were 42. Analyzing this does not fall within the get 10,000+ Twitter followers easily, quickly, and ethically time frame.

If you finish your daily ritual in fewer than 15 minutes, consider endearing yourself to others by directing a tweet to them with the @ sign. For instance:

Don't you think @verywell (circled above) was pleased when Darla Leal directed this tweet personally? If @verywell wasn't already following Darla, he or she probably did after seeing this.

A really great way to get followers by having them notice you is to include hashtags for a particular topic in your tweet, as TokenPet did here:

Token Pet's endearing tweet will appear in the streams of #dog, #cat, #doglover, #catlover, #animal, #psychology, #petlovers, and #pets. And so everyone targeted will see this tweet and instantly love Token Pet and decide to follow them. I ~~guarantee~~ am sure of it. You can have similar results by crafting your own lovely message or image, listing some hashtags, and pressing "Tweet." But, as mentioned before, go easy on the number of hashtags you include in each tweet; don't look "soupy." Try posting the same tweet at different times with different hashtags. I recommend no more than three hashtags per tweet.

Like (click on the little heart) as many tweets as you want—it's a great way to get people to notice you and probably follow you back without messing with Twitter's algorithm. But be careful about retweeting. If you do a lot of retweets and don't include someone's handle (i.e., @CathyTurneyLafs), Twitter will think you're spamming and clip your wings. And when you click on the retweet symbol (circled below)

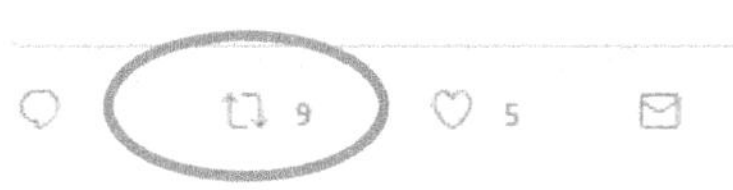

Do NOT fail to add a comment when this box pops up:

If you don't make a comment and you direct the retweet to someone (e.g., @CathyTurneyLafs) Twitter will, again, think you're a robot and there may be consequences.

Connecting These Dots

So, in summary, what was Part III, "Soaring," all about? Simply put, fun and productive ways to speed you to collecting 10,000+ Twitter followers easily, quickly, and ethically:

- How to create your own tweets that sing
- Best practices for posting and following
- Creative little ways to circumvent Twitter's annoying algorithms and add even more followers
- Clustering your not-to-be missed tweeters into lists
- Maintaining balance
- Hashtags—they're simply invitations to parties
- Best days to post

I suggest that if Part III is TMI (too much information) at this point, ignore it! Truly! I didn't know any of Part III's enhancements when I launched myself into the Twittersphere, and I gathered a significant flock just by doing what you read in Parts I and II. Get the basics down first—maybe accumulate 5,000 followers and then come back to Part III. By that time you'll be so adept at the routine that trying something new will be easier and more fun.

There's lots written about Twitter, with the effect of totally confusing wannabe tweeters. The information on the pages of this book is all you need to get 10,000+ Twitter followers easily, quickly, and ethically. Focus! Don't feel overwhelmed. Fear not! You can't break Twitter, and if you say something that isn't well-received remember that the public has a very short ~~attention span~~ memory.

So come down from that yoga headstand and have fun!

Before I Fly Off

If you're just beginning to write a book or you have a product in development, think how much more relevant it (and you) will be when at launch time you have 10,000+ Twitter followers that you amassed easily, quickly, and ethically. Don't wait until your product or book launches to start!

Twitter is essential today if you need to:

- Sell a product or service
- Look relevant
- Stay informed

There is no other platform that accomplishes this as quickly, easily, and inexpensively and without the limitations and restrictions that its competitors impose. Having a presence and many followers on Twitter tells the world that you have something to say that others want to hear. It's not hard.

Stake your claim to Twitter real estate now. It will be one of the best investments you ever make in getting into people's minds, hearts, and pocketbooks.

How do you eat a giant flying pterosaur? One bite at a time. Enjoy!

About the Author, and…

CATHY TURNEY is an avowed technophobe who knew that if she didn't conquer Twitter, it would conquer her. She is also an award-winning humor journalist and recipient of the American Business Awards Gold Stevie (Best Business Book of the Year) for her book *Laugh Your Way to Real Estate Sales Success* which she promoted to No. 1 in its category on Amazon by using Twitter. Cathy is managing partner at Better Homes Realty in Walnut Creek, California.

Upon learning that it would cost hundreds of dollars a month for someone else to set up and maintain her social media platform to market her first book, she had no choice; Cathy sweated bullets making Twitter *just work* and evangelizes about the easy methods she discovered to understand Twitter and amass followers.

Thank you for purchasing this guide. If you found it helpful, Cathy would greatly appreciate your posting a review on Amazon and Goodreads.

Sharing is caring. If you know someone who could benefit from this book, please tell them about it.

Cathy loves to hear from her readers. Connect with her at:

Email: Cathy@CathyTurney.com

Website: www.CathyTurneyWrites.com

Twitter: https://twitter.com/CathyTurneyLafs

Facebook: https://www.facebook.com/cathy.turney.3

LinkedIn: http://bit.ly/2lxgfqq

Pinterest: http://bit.ly/2oxSMUk

Goodreads: http://bit.ly/2sEIl7k

Acknowledgments

Acknowledgments…I could write a book. But it would need to be a multivolume set because my gratitude to these people runs so deep:

Lisa Tener. The award-winning goddess of voice who so "gets" what her authors mean to communicate and molds their words into text that sings. (https://www.lisatener.com/)

Rusty Shelton. Best-selling author and founder of the award-winning marketing agency Shelton Interactive who, in five minutes, overcame my resistance to Twitter and launched me into the Twittersphere…when all others had failed. (http://rustyshelton.com/)

Cynthia Rubin. Aka BestEditorEver, who shifts her focus from all her other editing to make brilliant commentary on my writing and save me from public humiliation.

Frances Caballo. The guru of social media who keeps me and thousands of other writers up-to-date in the ever-changing world of ~~S&M~~ social media. (http://francescaballo.com/)

Michael Larsen. Cofounder of the San Francisco Writers Conference. Through whom I met the above-mentioned giants of industry. And who so generously shares his publishing clairvoyance with us, his disciples. (http://www.larsenpomada.com/michael-larsen/)

Brenda Knight and Kate Farrell. Awesome presidents of the Women's National Book Association–San Francisco Chapter, who redirected my focus from writing my next real estate book to writing about how to get 10,000+ Twitter followers because "that's what we *really* want to hear about!" Bless you, ladies. (http://wnba-sfchapter.org/)

Paul Lander. Award-winning producer and writer for Showtime, Disney Channel, National Lampoon…(the list—it doesn't end). You are a comedic genius and the dearest friend. Thanks for honoring me by writing the foreword to this book. (http://www.humortimes.com/author/plander/)

Ron Childress. Aka WebWizard, who plumbs the depths of my mind to decipher my thoughts about what I *think* I need and translates them into a far better product than I could ever have imagined. (http://smmcats.com)

Howard VanEs. Amazon genius and master of all things Kindle. http://www.letswritebooks.net)

Angie Viets. Licensed clinical psychotherapist and author who uses her voice to heal people with eating disorders. Thank you for your amazing insights in your beta read of my book. You rock! (https://www.angieviets.com/)

Jeanette Zalfini. Dear friend and "enabler"! Thank you for being such a great sport about beta reading.

John Turney. MyHusbandTheEngineer…I could write a series. God, how I love you.

And speaking of **God**: My eternal gratitude.

Made in the USA
Coppell, TX
19 May 2020

25900440R10057